THE VISUAL ELEMENTS—DESIGN

THE VISUAL ELEMENTS SERIES

In these essential handbooks, research scientist and award-winning photographer Felice C. Frankel offers accessible guidance for scientists and engineers who must communicate their work visually for grant applications, journal submissions, and conference or poster presentations. The format of these books—each focusing on a crucial aspect of visual communication—is new, but Frankel's goal is not. Over the past twenty-five years, in her writing, popular online courses, and public lectures, Frankel has shown scientists and engineers the importance of presenting their work in clear, concise, and appealing ways that maintain scientific integrity. When she helps researchers create beautiful images and graphics of scientific phenomena, she is interested in more than helping them reach their research community or gain public attention. Frankel shows that the right visual elements also offer the power of reflection—images that help researchers look longer and understand more fully their own work.

ALSO PUBLISHED IN THE SERIES

The Visual Elements—Photography

The Visual Elements
Design

A Handbook for
Communicating Science
and Engineering

FELICE C. FRANKEL

The University of Chicago Press

Chicago and London

The University of Chicago Press, Chicago 60637
The University of Chicago Press, Ltd., London
© 2024 by Felice C. Frankel
Published 2024
Printed in China

33 32 31 30 29 28 27 26 25 24   1 2 3 4 5

ISBN-13: 978-0-226-82916-6 (paper)
ISBN-13: 978-0-226-82917-3 (e-book)
DOI: https://doi.org/10.7208/chicago/9780226829173.001.0001

Library of Congress Cataloging-in-Publication Data
Names: Frankel, Felice, author. | Frankel, Felice. Visual elements.
Title: The visual elements—design : a handbook for
    communicating science and engineering / Felice C. Frankel.
Description: Chicago ; London : The University of Chicago Press,
    2024. | Series: The visual elements
Identifiers: LCCN 2023024886 | ISBN 9780226829166
    (paperback) | ISBN 9780226829173 (ebook)
Subjects: LCSH: Visual communication in science—Handbooks,
    manuals, etc. | Graphic arts—Handbooks, manuals, etc.
Classification: LCC Q223 .F728 2024 | DDC 502.2—dc23/
    eng/20230928
LC record available at https://lccn.loc.gov/2023024886

♾ This paper meets the requirements of ANSI/NISO Z39.48-1992
(Permanence of Paper).

Contents

Introduction

This book is about design. But let me start by explaining what I mean by "design." In these chapters, you will not learn how to design an experiment, nor will you learn how to design a new device or find engineering inspiration while observing nature. These are important skills, but by "design," I mean arranging text and visual representations to communicate.

I am convinced that knowing how to design a figure, a journal cover, and a poster or slide presentation should be part of every researcher's education. A well-designed visual representation

- helps attract attention to your work,
- connects you to your research community,
- encourages collaboration,
- helps with funding,
- makes science accessible to the public, and
- clarifies your thinking about your work.

I want to emphasize this last goal, clarifying your thinking. In my twenty-five years working with researchers to help them communicate, my biggest challenge has been to curb researchers' desire to show it all—every image, every bit of data, and every graph your software can generate. I get it—you have done all the work, and you want to show it. But the truth is that your audience, even if they are in the same discipline, won't *see* it all. Graphics that include *all* your data are simply too much to look at and therefore too much to communicate. By trying to show everything, you end up showing nothing.

Take some time to think about how first-time viewers would see your graphic. What is the most important thing you want to convey? How can you help them see it? Can you summarize your research visually? If you can't, ask yourself if the goals of your research are clear in your own mind. In thinking about how to organize the visual elements of your graphic, you might even realize that you have not yet made a strong case for publication and need to do some additional research before your submission reaches the reviewers. (Maybe this book is about designing experiments after all!) Thinking about design, then, can sharpen your thinking about your science.

Design Is All around Us

I'll start with a nudge to get you into the right mindset. Look around you. Any time someone creates a visual for a sign, a map, or even a control panel, they make design decisions. A designer has to ask a number of questions: Who am I designing for? What is the first thing I want them to see? Can I create an order of things for them to look at—that is, a hierarchy of information? As a researcher, you should be asking the same questions when you create a figure for a journal submission or prepare a slide for a talk or poster.

Consider a visual that you've likely seen before (0.1). It depicts a concept about our solar system that was new in the sixteenth century. The astronomer Nicolaus Copernicus used it to illustrate heliocentrism—with the Sun, instead of Earth, at the center of the solar system and the planets orbiting around it. Although historians debate who actually read the book (or cared about the cosmology), the image from the book certainly has made an impact. I wonder whether school-aged children would

0.1

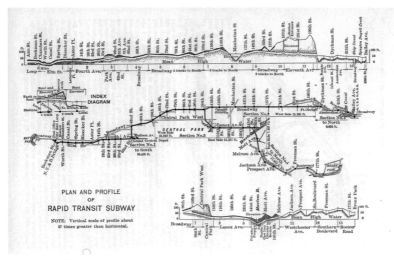

know the name Copernicus if he had used only words and no diagram. Would the concept be as clear and compelling without the clarity of those concentric circles, which show the relationships among the orbits?

Fast-forward to 1904, when designers created a map of the nascent New York City subway system (0.2). Note that they labeled the subway stops in the middle of the map and included additional maps of track elevation at the top and bottom. Were those necessary? Knowing the elevation probably isn't important for a subway rider, but it might be relevant to the engineers responsible for clearing snow off the tracks. Again, consider your audience as you design.

The designer of a map for the Boston Marathon also decided to include elevation at the bottom of the map (0.3). In this case, it's easy to understand how elevation is relevant to runners.

Here's another subway system map, this time of the Red Line in Cambridge, Massachusetts (0.4). I was taking the outbound train north, going toward Alewife from Kendall. The names of the stops before Kendall are not indicated. Why not? The designer understood that travelers heading in the opposite

0.3

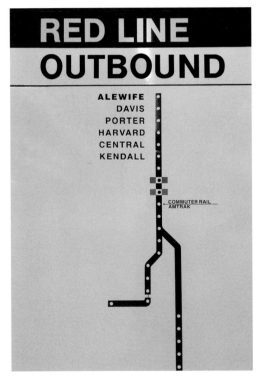

0.4

0.5

direction would not be interested in those previous stops. And in fact, it might be confusing—because those stops aren't options anymore from the platform. The design pushes viewers to focus on the relevant information. This was a thoughtful decision about what to leave out—a challenge for many of my students and collaborators.

When you step off the subway, you still can't escape design. On a quiet (at least for the moment) Boston street, a city employee is explaining visually where to place various components for a sidewalk renovation (0.5). Notice the decisions she made to clarify the design using color, lines, shapes, text, and arrows. We aren't her audience, but she is using a vocabulary familiar to those who will do the work. Note, for example, the repeated black shape under the word *concrete* and the pink diagonal lines, which suggest a border of some sort.

While driving in your car, you will see how someone made decisions about the colors to use for a dashboard map (0.6). Note the intuitive use of blue (for water), red (for one way or the wrong way), green for "go," and grayscale to show relative unimportance. I'll talk more about this later.

0.6

When you travel, try to notice how smart graphic designers give directions or show prohibited activities without the use of words (0.7). These abstractions are clever and effective because we quickly get the point.

0.7

The Origin of the Solar System Elements

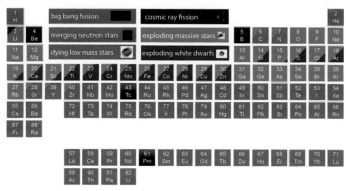

0.8

What to include and what to leave out is a question you should continually ask yourself. In another example, the chemist Jennifer Johnson redesigned the periodic table to *add* information about the origin of each element (0.8). She used color to indicate which elements were created in the big bang, with the merging of neutron stars, and during other cosmic events.

In contrast, a once widely distributed graphic left out important pieces of the story of evolution (0.9). Take a hard look at it. First, thankfully we do not all become white men. Second, where are all the species that did not survive? In other words, where is the story of extinction? That is the critical part of evolution, after all. Be sure to pay careful attention to what you decide to leave out in your own work.

0.9

What's Next

In this book, you will learn techniques to improve your graphics, and you'll see many examples of how others have put these techniques to good use. I've organized the book into four chapters.

1: LISTING AND SKETCHING

Chapter 1 covers formulating a list of components that must be included in your graphic and then sketching them, which can help you clarify your thinking and create a visual hierarchy that draws your attention to the elements that are most important.

2: CASE STUDIES—ITERATING THE ITERATIONS

Even experts encounter obstacles and dead ends. In this chapter, professional designers and researchers-turned-designers describe their iterations to create a compelling graphic, explaining where they started and where they wound up. You'll see fascinating illustrations from a variety of scientific disciplines. I encourage you to read through them all, even if the scientific discipline is different from your own.

3: GRAPHIC SUBMISSIONS—FIGURING THE FIGURES

This chapter discusses how considering color, composition, layering, and other design details can help you communicate more effectively by focusing viewers' attention on what's most important. I have included a wide selection of before-and-after figures to show how small changes can have big effects, as well as visual abstracts and thumbnail table of contents (or TOC) images.

4: POSTERS AND SLIDE PRESENTATIONS

This chapter discusses design considerations specific to posters and slide presentations, the two most common forms for presenting research findings other than journal submissions.

1

Listing and Sketching

Before you start work on any graphic, I encourage you to organize your thoughts and put them to paper—literally. First, write down a list of the visual elements you plan to include, distill them down to the very essential pieces, indicate their order of importance and then create sketches. Resist that urge to include everything just because you have the data or images.

Try working on paper before going to your favorite software, because that software will influence your thinking. The program's defaults push you toward certain layouts, which often do not benefit the communication of your research and its particularities.

For years I have had the pleasure of sitting down with researchers and listening to them explain their science. Often, they spontaneously take out a paper and pencil and start sketching to depict a concept. They clearly have an intuition for visual explanation. But somehow, when it comes to creating a graphic to submit to a journal, they fall back on software templates. The result tends to be generic-looking, even impenetrable, figures.

Your research is unique, so your figures should be too! Don't make your figures appear like everyone else's. Your figure should have its own "voice." Once again, think carefully about an order of importance, which might change from your initial thinking.

Try to see the relevance to your own work in the examples in this chapter, some of which come from outside science.

Making a List

A list doesn't have to be a boring string of words, especially if the list is intended for someone else. For example, while you may be familiar with Michelangelo's sketches, his list is something entirely different (1.1). I asked my good friend Mac DeVivo and her Italian colleague Lucio Miele to translate it. Mac wrote to me:

It's handwritten *volgare* (early Italian), and his handwriting wasn't easy to decipher. It turns out it's a GROCERY LIST. I found the original document, which is preserved in the Buonarroti House museum in Firenze. It contains instructions to buy certain items at the market, presumably for a servant who was illiterate, because next to each item there is a little sketch of what it would look like. What's interesting is that some of the items are repeated more than once (maybe the groceries were going to different

1.1

people?) Bread is listed three times, herring twice, wine twice (once by name, Lambrusco, the other generically—maybe cheaper house wine?) and tortellini twice. It's interesting that tortellini were around in 1518, the date of the letter on the back of which the grocery list was found!

Michelangelo put considerable thought into making his list. He clearly thought about how best to communicate visually with someone who couldn't read. But at the same time, he made sure the important bits were not left out: don't forget the tortellini!

One nice thing about lists is that you do not have to be an artist to make one. Here was my attempt as a nonartist to create a chronological "list," or a timeline, depicting milestones of a series of conferences and workshops I founded for an international group of scientists and engineers (1.2). The purpose of the conference series was to advance the visual expression of science, and I made the timeline to assist with a grant evaluation by our funding agency, the National Science Foundation. There is no question that making the document helped me to organize my own thinking and to make my own judgments about whether we were on the right track. For this sketch, *I* was the audience, and the timeline was intended for me. Using dashes, for example, for "preconference" activities versus solid lines for after the conference, helped me quickly see what happened when, so I could then communicate the process to the funding agency. When I wrote my final proposal for the funding agency,

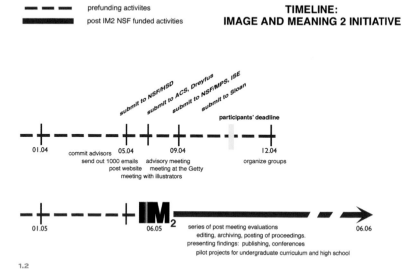

1.2

Visual Budget

Attendee information packet

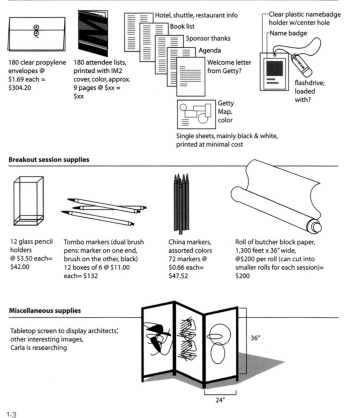

180 clear propylene envelopes @ $1.69 each = $304.20

180 attendee lists, printed with IM2 cover, color, approx. 9 pages @ $xx = $xx

Hotel, shuttle, restaurant info
Book list
Sponsor thanks
Agenda
Welcome letter from Getty?

Getty Map, color

Single sheets, mainly black & white, printed at minimal cost

Clear plastic namebadge holder w/center hole
Name badge
flashdrive; loaded with?

Breakout session supplies

12 glass pencil holders @ $3.50 each= $42.00

Tombo markers (dual brush pens: marker on one end, brush on the other, black) 12 boxes of 6 @ $11.00 each= $132

China markers, assorted colors 72 markers @ $0.66 each= $47.52

Roll of butcher block paper, 1,300 feet x 36" wide, @$200 per roll (can cut into smaller rolls for each session)= $200

Miscellaneous supplies

Tabletop screen to display architects', other interesting images, Carla is researching

36"

24"

1.3

I decided to include the sketch. I realized that the "tool," first intended just for me, would be a powerful visual to explain the project. By the way, we got the grant.

Lists don't necessarily have to be quite so linear. The talented science graphic artist Rebecca Perry—at the time working toward a graduate degree in science and technology—organized a visual list of items we would need for one of the Image and Meaning conferences to help us begin to see, literally, what our budget would have to include (1.3).

Making a Sketch

Sketches offer more flexibility than software, and they have a long history in both art and science. Leonardo da Vinci's volumes of sketches were a means for him to study the complexity of the human body in depth. His sketches of hand gestures helped him experiment and clarify which gestures evoked the right mood in his final masterpieces (1.4). They were studies in every sense of the word.

1.4

The French painter Edgar Degas did something similar. Look at the legs of the dancer at the right side of one of his sketches (1.5). Degas sketched alternative positions for both legs. The act of sketching expanded his thinking to consider different possibilities and even change his mind—an interesting concept, don't you think?

1.5

1.6a

1.6b

Today, technology enables us to study hidden sketches of the old masters, such as figures painted in a base layer that were subsequently painted over, invisible to the naked eye in the final painting. We've discovered how some great artists changed their minds when creating famous works of art. Sketching and resketching were part of their thought process. Consider a detail from an underpainting in Jan van Eyck's *Arnolfini Portrait*. Note how he experimented with the angle of the hand (1.6a) before deciding on the final position to capture just the gesture he wanted (1.6b).

SKETCHES TO COMMUNICATE STRUCTURE

In thinking about structure, consider a seventeenth-century illustration created by the natural philosopher Robert Hooke (1.7). The illustration shows the eyes and head of a drone fly, observed under a microscope. What's missing? The fly's body,

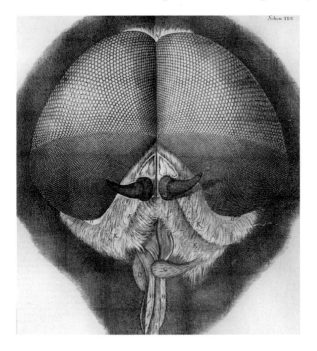

1.7

of course. Hooke also left out bits and pieces of material like dust or even certain bits of the fly's anatomy that were not exactly neat and tidy. Remember to think about what to leave out as you sketch—this is good practice for when you begin to create your figures. However, always keep in mind to maintain the integrity of the science in your editing decisions. I discuss this in depth in the *Photography* handbook.

The mechanical engineer Audrey Bosquet initially planned to use a photograph of the apparatus she used to analyze material for sports equipment, but she realized the photograph was too busy (1.8a). Instead, she used the photograph as a template for a wonderful hand-drawn sketch, which is so much more interesting and uniform (1.8b).

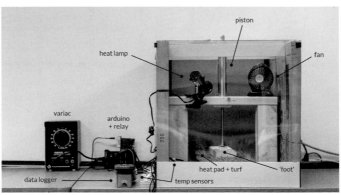

1.8a

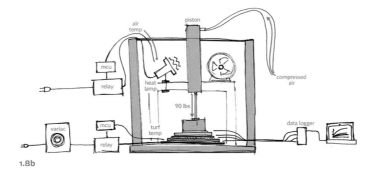

1.8b

Some scientists enjoy sketching by hand. The mechanical engineer Bavand Keshavarz drew the setup for his high-speed photography of falling drops (1.9; the original image has a black background, which I inverted for better reproduction in this handbook). Photographing the setup presented lighting challenges, where the hand-drawn look is quite appealing and clearly shows the main components.

Combining a sketch with a photograph is another useful strategy. The materials scientist Polina Anikeeva, who works on devices that interact with the nervous system, also enjoys sketching. She combined a sketch she made of the human spine with a photograph by her collaborator Alice Lu of optical fibers used in her research (1.10).

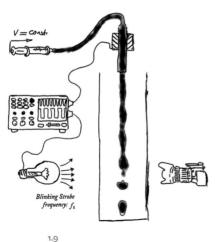

Strobing Dancing Jets

1.9

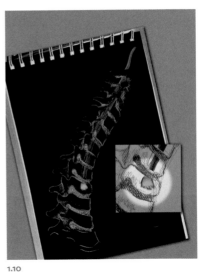

1.10

In another example, the mechanical engineer Alice Nasto decided that the photograph of her experimental setup had too much going on (1.11a). Like Audrey, Alice used the photograph as a visual template to make an illustration, deleting extraneous information, this time using software (1.11b). The result is a cleaner, simpler, easy-to-understand digital sketch.

1.11a

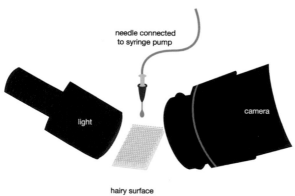

1.11b

1.12a

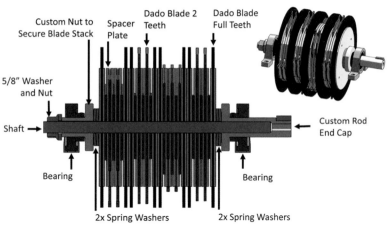

1.12b

Likewise, the mechanical engineer Sarah Southerland first made a photograph of the axle she was studying (1.12a) and then used computer-aided design software (CAD is drafting software used by architects and engineers) for her final, more readable depiction (1.12b). Note the easy-to-understand labels.

Sometimes sketching the structure of what you're studying can lead to a conceptual change. During the 1880s, the Spanish anatomist Santiago Ramón y Cajal made meticulous sketches that became the foundation of one of the most central ideas in neuroscience: the neuron doctrine, which holds that the nervous system is composed of discrete cells known as neurons. Notice the gaps between cells in Cajal's drawing of the spinal cord (1.13a).

1.13a

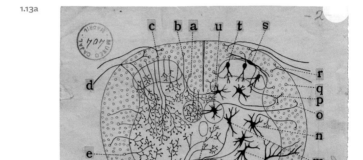

1.13b

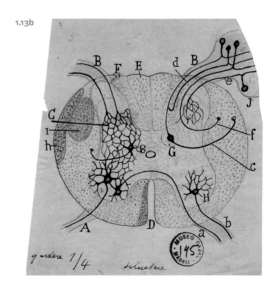

Cajal's work helped overturn the previously influential reticular theory, which stated that the nervous system was a continuous network—analogous to the circulatory system—with no separation between cells. Cajal illustrated this in another spinal cord drawing as well (1.13b).

A few decades earlier, Charles Darwin drew for himself a sketch of an irregular branching tree in order to make sense of his thoughts and observations on evolution (1.14). The making of the sketch was such a critical part of this thinking that it was the only sketch he included in *On the Origin of Species*. Note the important words written at the top: "I think." Sketching clarified his thinking for a game-changing concept.

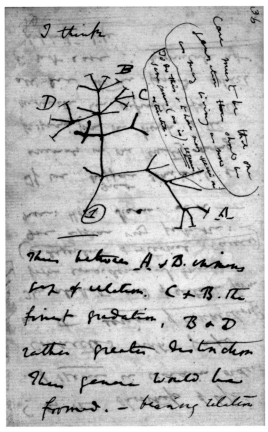

1.14

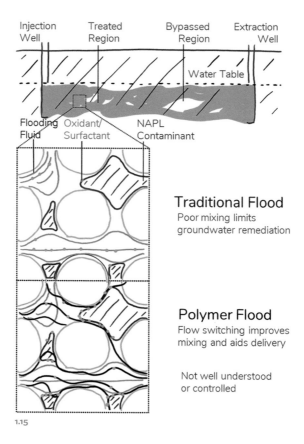

1.15

Here's a more modern, considerably less known example (1.15). Chris Browne made a sketch for his PhD thesis in biological and chemical engineering. It shows an approach to groundwater remediation. Chris drew it with a touchscreen-enabled pen—but didn't use premade illustrations from software—which gives his illustration a more personal quality.

THE SKETCH AS METAPHOR
If you think about it, visually communicating a concept that cannot be imaged directly usually starts with a sketch, whether it be

on paper or just as a thought. Today, we often see such concepts on the covers of journals. The bioengineer Omid Veiseh and the chemist Arturo Vegas asked me to create an image to submit for a *Nature Medicine* cover. They first sketched the concept of their work—a visual metaphor for an experimental approach to treating diabetes—on a whiteboard (1.16a). The research involved encapsulating transplanted pancreatic cells in compounds derived from algae to protect them from being destroyed by the host's immune system. I then put together bits and pieces of photographs to suggest the visual metaphor they requested— a needle and a drop of cells prepared for transplantation (1.16b).

Another time, I met with the mechanical engineer Jeehwan Kim to discuss a cover submission to *Nature*. Kelly Krause, *Nature*'s creative director, first sent me a quick sketch of what she had in mind—a visual metaphor for remote epitaxy, a process for layering thin films that has potential applications in making flexible electronics and other devices (1.17a). The idea would

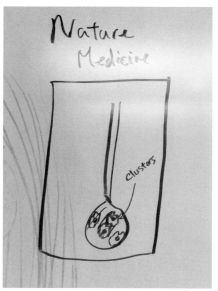

1.16a

1.16b

be to capture the basic idea of the process, which again can't be captured with a camera. After a further exchange, we agreed on some images that we could use as components (1.17b), and Nik Spencer, also on the *Nature* team, did his magic and finalized the image (1.17c).

1.17a

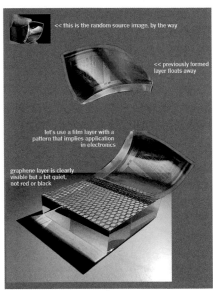

<< this is the random source image, by the way

<< previously formed layer floats away

let's use a film layer with a pattern that implies application in electronics

graphene layer is clearly visible but a bit quiet, not red or black

1.17b

1.17c

Some time ago, I ran a five-year National Science Foundation program called Picturing to Learn. My team of educators, managers, and faculty collected more than two thousand drawings from university students. We asked students to create sketches to explain various concepts they were learning in their classes to a high school student. Faculty gave them a list of basic principles, and each sketch had to include at least three principles.

One student made a sketch to explain the concept of Brownian motion—the random movement of particles suspended in a fluid and how that movement is affected by an increase in temperature (1.18). He used bumper cars as a visual metaphor for the particles, which made for an engaging sketch, but unfortunately he incorporated only one basic principle. The biggest problem is that the atoms or molecules that make up the liquid aren't depicted:

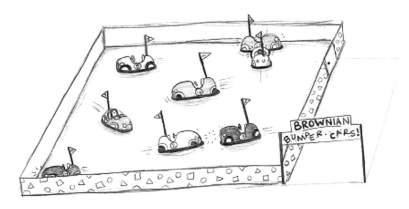

* more energy/speed [caused by higher temperature] causes more collisions & more intense chaos!

1.18

- ✗ Particles are different sizes.
- ✓ Temperature affects the rate of particles' movement.
- ✗ Atoms or molecules that make up the liquid are in constant thermal motion; their velocity distribution (or mean kinetic energy) is determined by the temperature of the system.
- ✗ Those moving atoms or molecules strike the suspended particles at random, making them move randomly through the fluid.
- ✗ Particles are suspended in fluid and are larger than particles of the fluid.

After evaluating all the drawings, we concluded that the vast majority of students had misconceptions about the phenomena they'd been asked to explain. In other words, sketching revealed the flaws in their understanding.

Some did meet the requirements. Another student's sketch explains why electrical current easily flows through conductors but not insulators (1.19). Here, a cliff represents the energy gap that blocks the passage of electrons (depicted as would-be party-goers) in an insulator.

1.19

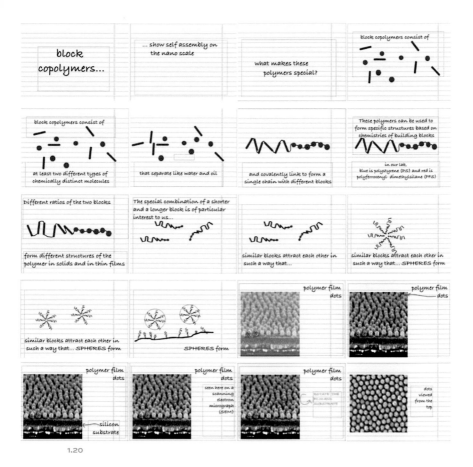

1.20

In another part of the program, we asked students to sketch storyboards for an animation that would explain a particular science concept. One student drew a storyboard to explain how molecules called block copolymers self-assemble—a potentially useful process for fabricating nanodevices (1.20).

After sketching the main concepts, the student met with me several times to develop the animation. At a certain point, we met with the researcher on whose work the animation would be based. In her final report, the student wrote: "This last step was crucial in the learning process: coming to a teacher with sketches and a visualization of what I thought the science meant made it possible for the professor to see where my thinking was wrong and what had to be explained differently. The animation became an interface for learning." The point is that the act of sketching is a means of evaluating your own understanding of the science. Think about that as you consider the case studies in the next two chapters.

2

Case Studies

Iterating the Iterations

I learned how to cook by watching Julia Child do her thing years ago. The best part of her television shows was when she made mistakes and then tried to correct them. I have always been drawn to understanding process—to the thinking behind the decision-making. That's what this chapter is about. Iterations let us think alongside design experts and scientists-turned-designers who are passionate about creating visual explanations. In the case studies in this chapter, you will hear from them directly, in their own words. I am immensely grateful to them all for their contributions.

You will notice in this chapter—and throughout the book—that there are few hard-and-fast rules. The "right" decision is specific to each exercise. Often you can see how different designers experiment with what to change, and then you can also see how they consider whether the change is an improvement. With your own design, if you aren't sure that a choice has improved your work, ask others inside and outside your lab. Remember, one goal is to build community and to communicate your ideas with others, so asking for opinions is helpful.

In these cases, I encourage you to navigate through the creators' iterations as they explain what they did and why they did it. When it's time for you to figure out how to tell your own story visually, you will remember that it is worth the time to consider more than your first attempt.

I have grouped these studies into three categories (the categories are not meant to be definitive): structural, conceptual, and numerical. You might argue that some of the cases could certainly fit in multiple categories, but I do consider the groupings useful for thinking through the creators' choices.

In the first instance, the examples distill physical things to their critical structural pieces and establish a hierarchy among those pieces. Once again, we make decisions about what to include and what to leave out—in the end, we are visualizing something. Put another way, we have whittled away the unnecessary parts that confuse the primary purpose of the illustration. In the second category, the examples give a physical property to a concept, process, or idea. In the third category, the examples give a physical form to a set of numbers. This category is all about data, but there is much more to say—more than I can say in this volume alone.

With all that in mind, let's take a look at some case studies.

Structural

In the examples that follow, you will see how the creators of each design took physical entities—a virus, a subway, an ocean bay—and decided how to represent their essential parts.

JONATHAN CORUM, *NEW YORK TIMES*

From day one of the COVID-19 pandemic, newspapers were flooded with all sorts of attempts to visually express the complexities of the structure of the virus along with the process of infection. In my opinion, Jonathan Corum's stunningly beautiful and clear depictions for a general audience are the best. We quickly get the overall structure of the coronavirus with an emphasis on the essential pieces of the virus that help it spread.

When the SARS-CoV-2 coronavirus began to spread around the world in early 2020, I started looking through scientific papers and other source material and thinking about how to draw it. Most real images of the coronavirus were fuzzy blobs, so my illustration would need to clarify the shapes of the prominent spikes—the critical proteins that enable the virus to infect human cells—and hopefully add a bit of personality. I tried to find a simple, modular design that would let me explain the different proteins in the outer layer of the virus and show the viral RNA coiled inside. These are the earliest sketches (2.1a).

Eleven spikes seemed like the right number to avoid bilateral symmetry (which can make drawings look too rigid, like snowflakes) and to not seem too crowded (in reality, each particle of the virus has twenty-four to forty spikes). One of the first widely circulated images of the coronavirus was an illustration with bright red spikes from the Centers for Disease Control and Prevention, so I used a similar red that might feel familiar to readers. The modular design helped me show how the virus opens to release RNA or falls apart on contact with soap (2.1b).

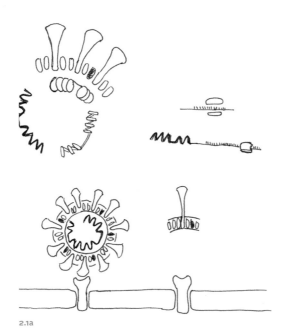

2.1a

2.1b

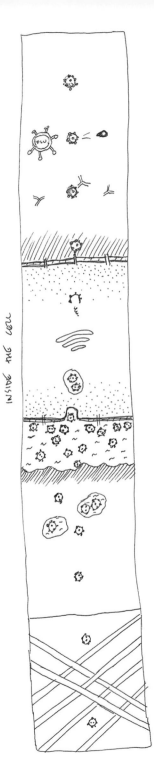

INSIDE THE CELL

2.1c

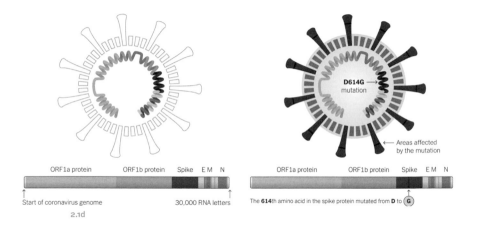

ORF1a protein   ORF1b protein   Spike   E M   N

Start of coronavirus genome   30,000 RNA letters

2.1d

The **614**th amino acid in the spike protein mutated from **D** to **G**

For my first large graphic on the coronavirus, I tried to explain how the virus enters and infects human cells. The rough sketch was fairly close to the finished graphic published online, which shows the process of a single virus entering a cell and then hijacking the cellular machinery to assemble new copies of itself (2.1c). The new copies of the virus then leave the cell and begin to spread.

For the rest of 2020, I tried to iterate on my original illustration and introduce something new each time. As scientists learned more about the coronavirus genome, I used color to visually link the coiled RNA inside the virus with a simplified horizontal diagram of the genome and its proteins (2.1d). The red segments, for example, encode the spike protein. Then, as individual mutations began to appear in the news, I used similar diagrams to locate each new mutation.

I was never happy with my early spikes—the tips felt pinched and spindly. Months later, after creating several three-dimensional models of the spike protein that ran in the *Times*, I redrew the spikes to be wider and more substantial (2.1e). My early illustrations of the virus also began to feel geometric and flat, so for variety, I tried roughening the edges and layering them like cut paper to create more textured shapes. But the original structure remained: the same eleven spikes at the same angles.

After a year of iterating on the same basic illustration, new coronavirus variants began to emerge with dozens of mutations each. I realized that my stubby horizontal genome diagrams

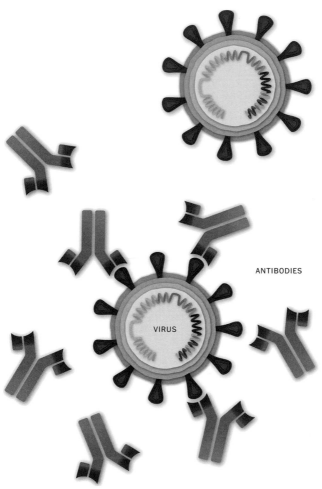

2.1e

would need to be much larger, so I tried wrapping them around the virus, mimicking the RNA coiled inside and letting me highlight each mutation with a line connecting the virus and its genome. The sharp lines made me think of butterflies pinned to a museum tray, so I paired the diagrams with world maps in the distinctive Cahill-Keyes projection to show where each variant had been detected (2.1f).

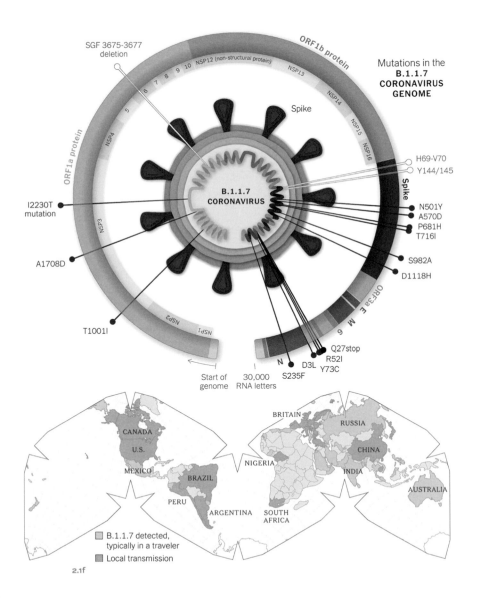

2.1f

As the coronavirus continued to mutate, a new naming convention (Alpha, Beta, Delta, Omicron) made it easier for the public to remember variant names. I continued to use these genome illustrations—in combination with more realistic 3D models—to explain each new variant.

ELLA MARUSHENKO AND KATE ZVORYKINA,
ELLA MARU STUDIO

Ella and her colleagues work in the world of cover submissions. As I know firsthand, sometimes you can think you've come up with a terrific image and it goes nowhere. Then, you experiment and make something better. Here, the back-and-forth conversation with the researchers is an essential part of the process.

This is an example of a *Nature Nanotechnology* cover image preparation. We were contacted by Jason White from the Connecticut Agricultural Experiment Station after the journal accepted his paper to discuss a potential cover design. The paper had quite a complex topic—not typically what the *Nature* editorial board chooses for the cover. The work became challenging and at the same time exciting.

The authors had developed copper-based nanomaterials and investigated their use for delivering nutrients and suppressing disease in plants. The paper focused on the new materials' design, but also on their potential application, so our challenge was to demonstrate both in the same image in an easy-to-understand way that was still scientifically accurate. Jason White and his colleagues sent us a graphic (2.2a). It shows the molecular structures of the designed materials as well as photographs of soybean plants treated with the materials in different stages of recovery from a fungal disease called soybean sudden-death syndrome.

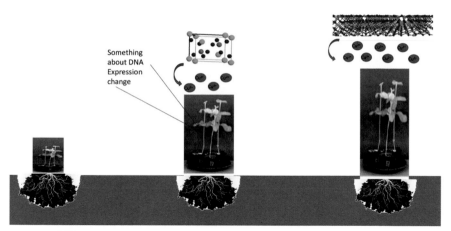

Something
about DNA
Expression
change

2.2a

The challenge was to depict different stages of recovery in plants with the help of nanomaterials. Taking a cue from the image sent by the authors, our first thought was to do this by depicting several plants and varying the size and quality of leaves to indicate the stage of recovery. We also wanted to show how the nanomaterials had altered gene expression in the plants, as the researchers had found.

Unfortunately, this did not work well in terms of composition and conveying the main idea. Therefore, we decided to depict only one plant but with different processes represented in different leaves. In the draft image, red leaves represent the diseased parts of the plant, and a single green leaf has been restored to health by the nanomaterials, which are shown as bright dots (2.2b).

We sent the first draft to the authors of the study and nervously waited for their comments because our idea differed from their first sketch. They really liked our concept but asked for several changes exaggerating our idea and making the details more pronounced. They asked to make the plant green (everything but the roots), with only one of the leaves wilted and sick and the roots partially black to indicate disease.

2.2b

They also asked for a number of additions, suggesting that we place some fungal spores around the roots to show the effects of the disease, represent the chemical structure of nanomaterials around the leaves instead of using the bright dots, and include an electron microscopy image of the leaf structure at one of the leaves. We were concerned that all these additions would make the image too complex and too much like a schematic to work well on the cover.

In the end, we changed the color of the plants so that most of the leaves were green and healthy and only one was sick, indicated by yellow and dark shading. We used bright dots to depict the nanomaterials as before, then added water molecules and ions to indicate the chemical reactions happening in the leaves. We made a tough decision to leave out the detailed molecular structures of nanomaterials because they would make the overall composition too cluttered. [This is a problem for those of us who submit cover ideas. Researchers simply want to say everything in just this one image!—*Felice*]

nature.com/nnano   December 2020 Vol. 15 No. 12

nature
nanotechnology

2D PEROVSKITES
Back to fundamentals
RAMAN MICROSCOPY
Single-shot particle localization
QUANTUM COMPUTING
A two-qubit register from nuclear spins

**Nanoscale copper for
crop nutrition and protection**

2.2c   2.2d

Two leaves near the bottom show details from the electron microscopy. They do not draw too much attention, yet they add some interest to the bottom part of the stem. The roots were the most complex part of the image because we had to add spores and show both healthy and sick roots, as well as the nanoparticles. We placed a few spores near one of the roots on the far left and made them semitransparent. These elements help balance the large leaves in the upper part of the image. We made the sick roots black and added copper nanoparticles as yellow dots, as earlier.

With these changes, we felt we had reached a good compromise between the extra detail the authors had requested and our desire to create an uncluttered image that still represented all the important elements (2.2c). We thought the image looked very nice, and the authors were happy, too. They sent it to the editors as a cover suggestion, and this image was chosen as the cover for the December 2020 issue of *Nature Nanotechnology* (2.2d).

DAVID S. GOODSELL, SCRIPPS RESEARCH
AND RCSB PROTEIN DATA BANK
*I've had the pleasure of knowing David for many years. His artwork
attracted attention when he was the first to show the complexity of
cells' inner workings. Here, he goes into depth, encouraging us to
spend time learning software so that we direct the software instead
of the software directing us.*

As a computational biologist, I am very lucky to work in a field
that has many powerful, free software packages for visualization.
Researchers have been refining these methods for decades, and
current tools give users a lot of freedom when creating new
images of molecular structures. Frankly, however, the default
images produced by these packages only rarely show what I want
to show. So, it's always worth taking the time to learn the various
options that are available for customizing images and using them
liberally to create an image that best displays the concept.

The example I'm using here is one of the most complex
atomic structures currently available in the Protein Data Bank, a
freely available online repository of protein structure data (2.3a).
It is the central portion of a flagellar motor, used by bacteria to
rotate long filaments that act like propellers when they swim.
This portion of the motor contains a central axle and several
ring-shaped bushings. The entire structure is composed of 219
protein chains. I made the illustration with software I created
called Illustrate. It required several hours to sort out because
I had to manually assign the colors of every chain. My painting
shows an artistic concept of the whole motor in the context of
the cell (2.3b).

Over the next few pages, I'll walk through some of the decisions that I made in this illustration, but using Mol*, the default
molecular viewer at the Protein Data Bank. If you want to try
this out, the Mol* view of the assembly is available at the RCSB
PDB (Research Collaboratory for Structural Bioinformatics Pro-

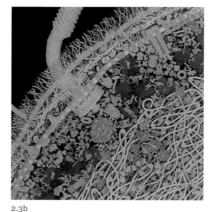

2.3a   2.3b

tein Data Bank) website for entry 7CGO. I used mostly simple point-and-click options available directly in the Mol* interface and generated all of the following illustrations in about fifteen minutes.

First, some thoughts about viewpoints. Most molecular viewers plop the molecule in the center of the viewer in an easy default orientation, in this case looking down on the long axis of the structure. This is never the view you want. Invariably, you have to reorient the view to highlight the interesting aspects of the structure. For this structure, the default view shows the symmetry of the lower ring nicely, but I rotated it about ninety degrees to be able to see the whole structure.

Also, most interactive viewers use a perspective view that enlarges closer parts of the structure relative to more distant parts. This looks very natural when you're turning it around interactively but can introduce misconceptions in static images. Compare the perspective view (middle panel) with the orthographic

DEFAULT VIEW   PERSPECTIVE VIEW   ORTHOGRAPHIC VIEW

2.3c

DEFAULT COLORING   POLYMER CHAIN ID   ENTITY ID

2.3d

view on the right (2.3c)—both are viewed from exactly the same direction relative to the structure. In the orthographic view, in which size doesn't vary with distance, it is obvious that the bottom ring is slightly tipped relative to the lower ring. In the perspective view, it's impossible to tell if the different alignments are real or simply due to the perspective transformation.

Then it's time to think about colors (2.3d). Most molecular viewers give many, many options for colors. By default, Mol* assigns a different color to each protein chain using a scheme that is designed to be visually differentiable (left panel). In this case, however, the default coloring scheme doesn't highlight the different structural components of the assembly—the central axle and the various rings—so I flipped through some of the hardwired options in Mol*. The "Polymer Chain ID" coloring (middle panel) assigns the color based on the order of chains in the data file. Because authors typically put all copies of a particular protein in a big block in the file, this scheme does a pretty good job, apart from the little color break on the right side of the lower ring. The "Entity ID" scheme (right panel) is better: it uses different colors for each *type* of chain, making it easy to see, for example, that the upper ring is composed of two types of protein chains (brown and pink). Notice that I've also tipped these two images slightly relative to the view of the default coloring, just to make the rings a bit prettier.

The "Entity ID" picture really shows the structural components of the assembly, but I'm personally not thrilled with the colors (2.3e). Fortunately, Mol* has an easy way to select all the copies of a particular type of chain and then use a point-and-click

2.3e

GAUSSIAN SURFACES   SPACE-FILLING REPRESENTATION WITH ALL ATOMS

2.3f

interface to change colors. This step is highly subjective, but at
the same time, it is loads of fun and can result in a highly personal approach to creating images. I tend to like clear, dynamic
colors, and I try to use similar colors to highlight the different
functional parts of an assembly. To create the image on the right,
I used blues and purples for the central axle, yellow and green for
the upper ring, and orange for the lower ring. I didn't try to tune
the colors at all—I just used the options that were directly available in the palette, so the whole thing only took a few minutes.
However, this can be endlessly refined (and I often do spend far
too much time playing with colors). In my original illustration,
for example, I colored every other subunit in the rings slightly
different colors to show that they are different chains but still
have them read as one ring-shaped assembly.

We also need to talk about representing finer structure.
Molecular viewers provide many options for displaying the
atoms in a molecular structure. All the pictures so far have used
a smooth Gaussian surface that shows the overall shape of the

proteins but omits all the details of the individual atoms. Alternatively, you can display all the atoms, as I did in the original illustration and in the Mol* version (2.3f). But for an assembly of this complexity, this makes for a very busy image. Other options, such as ball-and-stick and ribbon diagrams, end up being impossibly complex for structures this big, but they can be useful for up-close exploration.

Finally, most molecular graphics programs provide a bunch of tricks to help make the overall shape and form more interpretable. I usually play with these at the very end, since it involves fiddling with a bunch of parameters to get the best look. I used two fun options (2.3g). Most programs have some sort of "depth cueing" or "fog" that progressively changes the color of the molecule to the background color in distant regions. In my experience, this usually works best with black backgrounds (left panel), making it easier to understand depth relationships—look at the ragged edge of the axle at the top, for example, where the most distant subunits are darkened. Some programs can also add outlines to help us understand the shape. The version in the middle panel uses the same "matte" shading I've been using all along; the one on the right uses flat colors that produce a very cartoony image. I'd be happy to use any of these illustrations as an introductory overview figure in one of my publications!

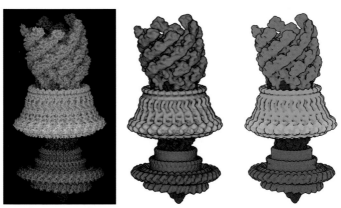

2.3g

GAËL MCGILL, FACULTY AND DIRECTOR OF
MOLECULAR VISUALIZATION AT HARVARD MEDICAL
SCHOOL AND COFOUNDER AND CEO OF DIGIZYME,
WITH JONATHAN KHAO AND GEOFFREY CHEUNG

Here, Gaël McGill and colleagues depict complex mechanisms of viral entry for a scientific audience by combining real structural data with judicious design choices. Gaël is writing for the scientific community and understands that he can go into more detail than Jonathan Corum, who is speaking to the nonexpert.

Many tools exist for visualizing molecules, from the ball-and-stick figures that adorn high school chemistry classes to sophisticated 3D models of proteins (see David Goodsell's case study). The challenge of molecular visualization, however, is amplified when one needs to depict molecules in motion—whether it's intermolecular movements (molecules moving around and interacting within a shared environment) or intramolecular motion (the changes in the conformation of a molecule that occur as it carries out its function).

The challenge is not just for novices but also a serious limitation for scientists who struggle to create dynamic models that integrate structural and dynamic data into a single model for inspection, hypothesis generation, and communication. A complex and, unfortunately, timely example of such a visualization challenge is the SARS-CoV-2 spike protein, a key part of the molecular machinery that enables the virus to infect human cells.

Like other coronaviruses, SARS-CoV-2 is enveloped by a membrane and delivers its infectious genomic payload by fusing its membrane with that of the host cell. The spike protein initiates fusion and undergoes drastic contortions in the process. Over the past two years, our team has created several animations and illustrations to help scientists and others better understand this complex process.

The spike protein has two parts: S1 and S2. The S1 portion of the spike is a protective cap that recognizes and binds to a receptor molecule known as ACE2 in the membrane of the host cell. Underneath S1, the S2 portion of the spike is a membrane-fusion machine that pulls the virus and cell membranes together. The spike also plays hide-and-seek with the host's immune system by disguising itself with a layer of flexible sugars called glycans that make it harder for antibodies to detect.

Upon binding ACE2 on a host cell, S1 falls off, exposing the S2 fusion machinery below. The molecular choreography that ensues is highly complex and involves the progressive unfolding and refolding of multiple parts of S2. These are the motions that provide the requisite force to drive the viral and cell membranes closer together and induce their fusion. What are the design choices that can most clearly introduce viewers to this protein and help explain this multistep molecular ballet?

For a flexible protein of this size and complexity, a surface representation felt like the right choice. This common type of molecular image sacrifices details of atomic structure but gives a good sense of a protein's overall shape. In this case, reducing the visual detail helps viewers get a broad sense of the spike's overall shape and how S1 and S2 fit together. This coarse surface representation was also well suited to highlighting the number and position of glycans (right panel in 2.4a). At this point, early in the animation, fewer colors and low tonal contrast in color groups helped keep the different parts of the spike clearly identifiable while the structure was moving. We used selective graying as a simple yet effective way to guide the viewer's attention to the features we wanted to highlight as the sequence plays.

Later in the animation, we wanted to convey a more detailed understanding of the spike's structure and flexible regions. This required that we shift to a backbone ribbon representation, a common way of depicting how the protein chain is folded in 3D

2.4a

space. Color also became a critical means of drawing attention to specific regions and types of data used to create the protein model. A series of images in three panels (from left to right in 2.4b) shows different uses of color to represent the spike's overall structure (S1 in red and S2 in yellow), the spike's detailed architecture (shown with multiple colors), and the provenance of the structural data used (red indicates regions lacking 3D structural data that were modeled with lower accuracy).

Once it came time to convey the underlying molecular logic of the spike's multiple refolding steps, a representation that shows details of the protein backbone was critical because it is the best way to appreciate how the protein "zippers up" parts of its structure as it folds. In another frame (2.4c), the virus (top) has lodged two spike proteins into the host cellular membrane (bottom) and is about to refold the spikes to pull the membranes closer together.

2.4b

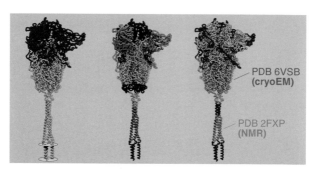

PDB 6VSB
(cryoEM)

PDB 2FXP
(NMR)

2.4c

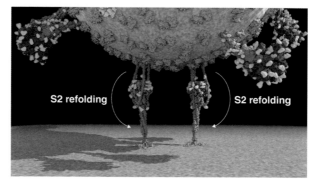

S2 refolding   S2 refolding

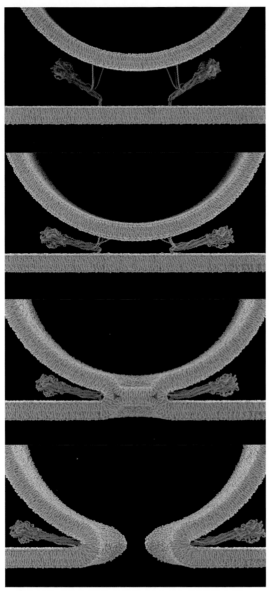

2.4d

Next, to reveal the molecular rearrangements that subsequently occur, we needed to switch to a cross-section view of the process and show individual lipid molecules, for which we had created a coarse-grained simulation. Panels show the sequence of stages in the membrane fusion process (2.4d).

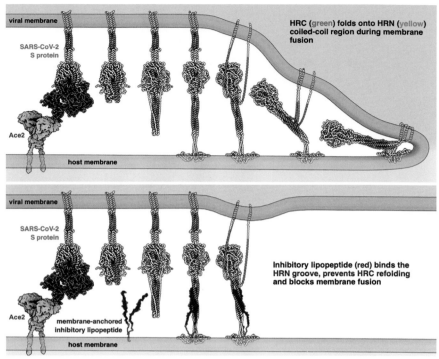

Finally, in an attempt to summarize this process and show how it could be inhibited by compounds called fusion peptide inhibitors, which are being tested as a potential therapy for COVID-19, we created a two-part layout (2.4e). The top panel shows key steps leading up to membrane fusion, and the bottom panel shows the same process blocked by a fusion peptide inhibitor tethered to the host cell membrane (in red). Because the peptide sequence of the inhibitor mimics a region of the virus spike protein, the inhibitor fits neatly into one of its grooves, thereby preventing other regions of the protein from refolding. This wrench in the cogs of membrane fusion can be understood only in the context of how the spike protein folds, and we hope this figure illuminates this unique mechanism.

JEANNIE PARK, COFOUNDER AND LEAD DESIGNER AT
DIGIZYME, WITH GEOFFREY CHEUNG AND GAËL MCGILL
*Jeannie Park takes us through some of the design challenges that she,
Geoff Cheung, and Gaël McGill faced to visually explain a challeng-
ing scientific concept, the Venturi Tube.*

Fluid dynamics is one of the most challenging areas of science to
understand and describe, especially when the target audience is
easily deterred by complex math. Yet fluids—the scientific defi-
nition of which includes both liquids and gases—are all around
us and touch every aspect of our lives, including the air we inhale
with every breath, the turbulent swirls of milk in our morning
coffee, and the flow of misty air around the wing of a plane as it
cuts through the clouds. All these phenomena, and many more,
require an understanding of how fluids behave and respond to
objects moving through them.

On a recent project about sustainable engineering based on
nature's models, our goal was to introduce, without math or the
need for extensive scientific background, the physics and engineering of life. One design challenge we faced was coming up
with a visual explanation of the Venturi effect: the increase in
speed and simultaneous reduction in pressure that occurs when a
fluid passes through a constricted section of pipe. This important
principle of fluid dynamics comes into play when explaining how
the shape of a wing achieves lift, enabling flight. The eighteenth-
century Italian physicist Giovanni Battista Venturi demonstrated
this in what is classically known as the Venturi tube experiment.

The setup is simple—an incompressible fluid moves through
a constriction in a pipe—but the effect is counterintuitive. When
the fluid enters the constriction, it increases speed, resulting in
a decrease in pressure. The mind too easily jumps to the analogy
of a traffic jam, where the opposite happens: as particles crowd
together, they slow down and presumably bump up against one
another, creating higher, not lower, pressure!

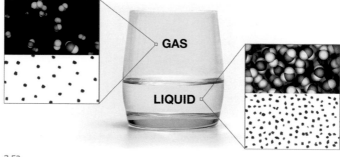

2.5a

To explain the Venturi effect, we needed to establish a visual language that could be applied consistently when demonstrating multiple concepts in fluid dynamics. To understand the nature of fluids, for example, we needed to explicitly define fluids both as a gas and as a liquid. In one illustration, we used a drinking glass half filled with water. As the gaseous form is easily overlooked, we reinforced this using magnified callouts (2.5a). Within each callout, water is represented both as molecules (red and white space-filling models of H_2O) in the top half and simplified as blue dots in the bottom half. By comparing the left and right representations of water as gas versus liquid in this way, viewers can easily see that water in both forms is made up of identical molecules but with a discernible difference in density. But this depiction does not capture an essential aspect that we will need for the Venturi tube: motion.

To imply movement and flow, it seemed logical to use dotted lines with arrowheads at the ends (2.5b). These would contrast with the solid lines of the obstacles we were going to use in some graphics to demonstrate how fluids move around solid objects. However, discerning the direction of flow in an area far from the arrowhead was problematic. We considered replacing each dot with an arrowhead but realized this could overcomplicate the image.

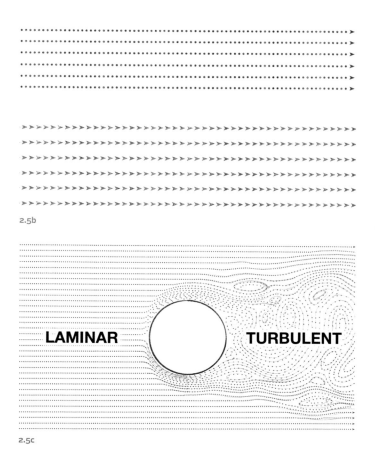

2.5b

2.5c

How could we effectively convey the direction, speed, and density of a fluid in the presence of an obstacle? For example, in representing smooth, organized "laminar flow" as dotted lines with single arrowheads at their ends and the turbulent motion resulting from an obstacle, we are able to distinguish organized and disorganized flow (2.5c). However, as the dotted lines in the turbulent area curl and twist, we risk losing a sense of direction with so few arrowheads. We underlaid faint lines beneath these dots to help give a better sense of flow and direction, but it's still difficult to get a clear sense of direction at any one point in the turbulence.

2.5d

The visual representation becomes further complicated when demonstrating how an object's size can affect how it moves through a fluid (2.5d). For example, a single-celled organism in a drop of water (symbolized by the small circle to the left) experiences resistance or drag within the fluid very differently than a whale swimming through the ocean (the oversized circle to the right). Even if the water is the same viscosity, the smallest object experiences water as a fluid as sticky as molasses, whereas the largest one appears to glide effortlessly and with conserved momentum.

To describe the movement of a fluid around such objects, one needs to come to terms with two competing mental models of fluids. On the one hand, fluids are an ensemble of molecules that can be visually represented by dots (as in our water-glass graphic). This representation highlights the "particle" nature of fluids and implies that they behave as a collection of individual molecules according to the rules of molecular mechanics. On the other hand, illustrating the principles of fluid dynamics that explain, for example, the magic of flight, requires one to step back and take a macroscopic perspective of how a fluid moves and interacts with other materials as a whole. On a macroscopic scale, fluids behave as a single substance, with properties like density, viscosity, pressure, and velocity.

We came to realize that if we used dotted lines to represent such interactions, one might easily confuse the dots with molecules. We decided to move away from any visual representations that would remind the viewer of a fluid's molecular composition and began to experiment with solid lines (2.5e).

But this brought challenges of its own. What are the right graphical choices to represent these key characteristics, such

2.5e

as speed, pressure, and incompressibility? Straight lines with arrowheads clearly suggest direction, but what do different line thicknesses imply? Does thicker indicate greater speed or larger volume of displacement to the viewer? What do the number of lines, their spacing or the frequency, and the shape of arrowheads communicate about motion? Finally, there was the issue of contrast and color. Can we effectively use light and dark contrast or multiple colors within the line to indicate speed, or would viewers confuse this with varying temperature or pressure?

With all these visual parameters in mind, we came up with a representation of the Venturi tube experiment (2.5f). The fluid is represented by "packets"—an intentional departure from any dot-shaped particulate representation—to remove any association or potential confusion with individual molecules. Darker shades of blue indicate higher pressure. In the graphical key, we suggest that pressure is four times higher in the wider portions of the tube than in the constricted area (based on Bernoulli's equation).

Venturi Tube Experiment

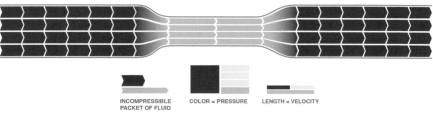

INCOMPRESSIBLE
PACKET OF FLUID

COLOR = PRESSURE

LENGTH = VELOCITY

2.5f

The length of the packets indicates velocity—the longer the packet, the faster it is moving. The key indicates that the fluid moves twice as fast in the constriction than in the wider ends of the tube. Finally, we altered the shape of each packet to suggest the direction of motion.

By making careful design choices, we hope to dispel confusion and establish a new mental model for the rather counterintuitive result of the Venturi tube experiment: the inverse relationship between pressure and speed of an incompressible fluid flowing through a constriction in a tube. The design choices that led to this set of visual elements not only effectively serve as the basis specifically for our project's discussion of flight and wing design but also establish visual cues that will be recognizable when reused to illustrate numerous principles of fluid dynamics.

JANE WANG, CORNELL UNIVERSITY

I had the pleasure of meeting Jane Wang during one of my online graphics workshops. She later showed me some fascinating videos, and we discussed the challenge of designing her video, capturing the "righting" reflex of a dragonfly, into a story in a still image. Here the original images (with black backgrounds) were inverted so that we could better see the process in print.

When I arrived in New York in the summer of 2021, I was unsure where I would find dragonflies, and the possibility of doing experiments in the city seemed remote. The next day as I stepped outside, a dragonfly flew by, reporting its existence, before quickly disappearing again. This started a summer of searching and filming dragonflies in the city.

When I returned to the lab, using a Photron 10 high-speed camera at one thousand frames per second, I focused on a dragonfly's action for research—how it rights itself when falling upside down. From a scientific point of view, the animal's righting behavior is an ideal experimental system for analyzing the interplay between the physics of flight and dragonflies' evolved neural algorithms. To quantify and explain their maneuvers, we design experiments, do mathematical modeling, extract the numbers from the images, and work with equations to make sense of the dynamics of the behavior.

We see images before we abstract them into numbers. We learn first by looking but spend increasingly less time looking these days because there is too much to look at. It is tempting and often necessary to ask computers to look for us. Making this video was intended for us to look and look again.

The video shows two righting maneuvers. The first was by design and the second was a surprise. Can you see them in the still (2.6a)? Toward the end of the sequence, the dragonfly bounced off the wall and found itself in an upside-down orientation again. It performed the same 180-degree roll recovery just like the first one. These unexpected collisions are one of life's uncertainties that forced dragonflies to practice their righting skills before evolving a hardwired reflex.

We then tracked the head, the tail, and the wings on the computer so that we could observe the body and wing motions during the recovery maneuver (2.6b).

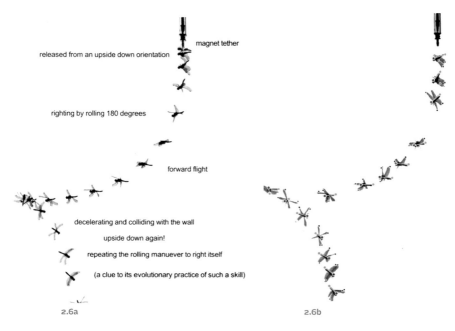

magnet tether

released from an upside down orientation

righting by rolling 180 degrees

forward flight

decelerating and colliding with the wall

upside down again!

repeating the rolling manuever to right itself

(a clue to its evolutionary practice of such a skill)

2.6a

2.6b

If we remove the dragonfly images, showing just the traces of the body and wing markers, viewers step out of the body frame and see the trajectories of the body and wings in space. Showing these traces on their own has a surprising visual effect. The rhythm of the flight is captured in these abstract lines.

In another image, each color traces the tip of one of the dragonfly's four wings (2.6c). And we can do the same with the head (red) and tail (green) (2.6d).

Combining the previous two images produces another (2.6e), showing the trajectory of the wing tips and the body.

This sequence of visual images in a way deconstructs how we see. We first notice the contour of a dragonfly, then the head and tail, then the trace of the wings. Step by step, we build an understanding of their movement in air. We learn by the process of seeing.

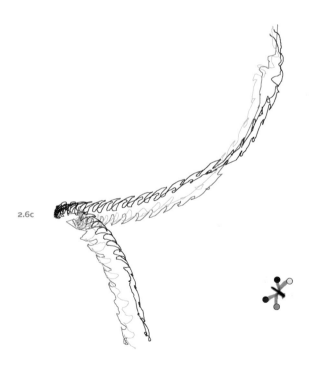

2.6c

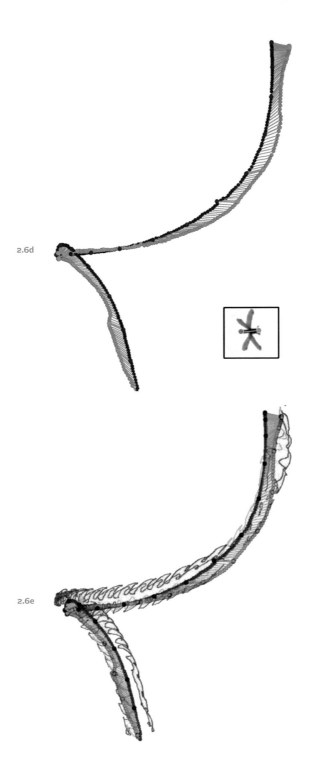

2.6d

2.6e

I became fascinated with the various attempts at representing the New York City subway system as it became more complex over time. The following is more of a brief visual history rather than a single iterative process of how a representation had to address the challenge of juxtaposing two discrete systems—the subway map and street map for users of the subway system. The issue of representing a thing that is constantly evolving is very much relevant in science. Think of the various changes that had to be made when visually describing the structure of an atom. I asked Eddie, who worked on a solution to the subway map challenge, to give an overview and to include his own thinking for his solution.

An underground transit system is, by its nature, hidden, or virtual, to the user. A transit map represents the public's reality of a transit system. A clear and accurate transit map is the critical link between the city and the transit system that serves it. It instills confidence in users, empowering them to venture underground to travel quickly and efficiently to their destinations.

Clear and accurate mapping of New York City's subway system has been a major challenge from its beginning in 1904 right up to today. First of all, it is one of the most complex transit systems in the world, with twenty-six separate lines serving 468 stations, most running twenty-four hours, seven days a week. Additional complicating factors are unique to New York City. First, the central business and tourist district, Manhattan, is a narrow island crammed with twenty-two subway lines. In addition, the subway uses an innovative system of "express" trains that alternate between express (skipping stops) and local (stopping at every station) and run alongside the regular local-only train routes.

For over half a century, New York's subway maps have been stuck in one of two distinct mapping categories, Geographic and Diagram, and both types are still used.

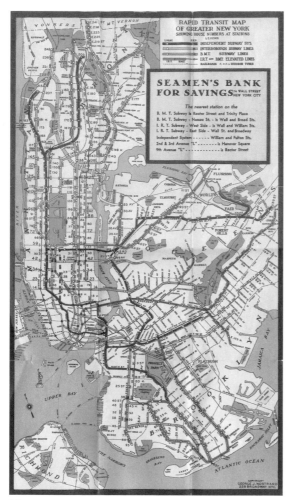

2.7a

Consider an example, from 1939, of a geographic map (2.7a). The approach overlays the subway routes on a street map. It was done back in the nineteenth century and is still used today. The deficiency of geographic subway maps is that they don't clearly show enough system information, for example, the fact that many of New York's express trains skip stations yet travel alongside the local trains. Figuring out which train stops where is difficult. In addition, the geographic maps show dense aboveground information, distracting the user from the subway routes.

A diagrammatic Vignelli from 1972 simplifies the subway routes and connections with lines that intersect at ninety- and forty-five-degree angles and might appear to be more "readable," but there is no clear relationship between the subway system and the city's aboveground streets or landmarks (2.7b). [For years the pros and cons of the Vignelli design have been argued. The book *The New York Subway Map Debate at Cooper Union April 20, 7:30 pm, 1978* is a transcript of just one evening of arguments.—*Felice*] It leaves out almost all street information to show the subway lines. Geographic accuracy is compromised, and without aboveground reference points, users can become disoriented.

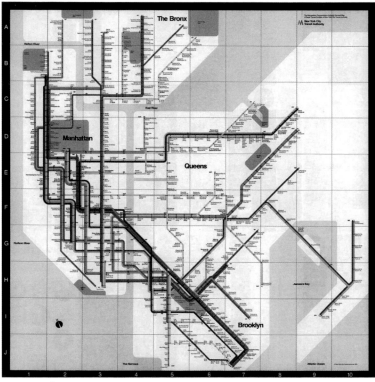

2.7b

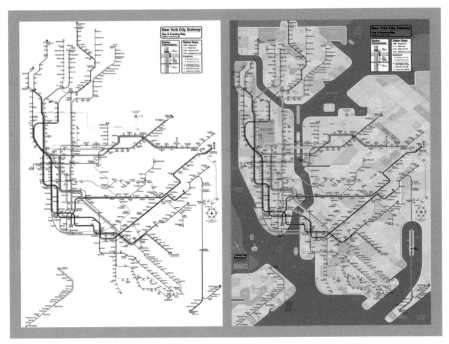

2.7c

Here is a newer attempt at simplifying the information (2.7c). My solution is the KickMap—a hybrid diagram that utilizes the strengths of both a diagram-based map and a geographic map. The subway lines are layered onto the fabric of the city—its streets, parks, and neighborhoods. On the left are just the subway lines without the New York City map background. It looks very similar to a standard twentieth-century style diagram, but it is not. It is carefully shaped and contoured to follow the major streets, landmarks, and parks of New York City. The map on the right has that diagram from the left layered on top of a stylized geographic New York City map. Hence *hybrid*—it combines the best of both diagram and geographic mapping. The hybrid diagram stylizes both the street map and the subway map so that they seamlessly complement each other. For users, this dual stylization accurately links the subway stations to relevant aboveground locations, eliminating confusion. Consider a detail of the hybrid, too (2.7d).

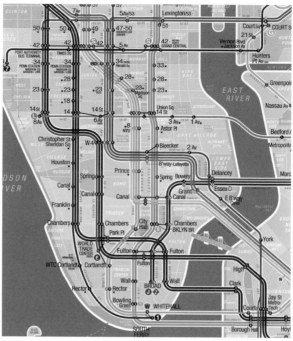

2.7d

The hybrid has been translated into a dynamic iPhone app (2.7e). One can touch and linger on any station to bring up a street map of that area to include pertinent information.

2.7e

Glenda is the senior illustration editor at Annual Reviews *(AR).*
I met her through Jennifer Jongsma, the associate editor in chief and
director of production at AR. Annual Reviews *has always under-*
stood the importance of visual representation and has budgeted for
a talented team of graphic artists to clarify the visual submissions of
the researchers.

At *Annual Reviews*, we often work on graphs or charts that are abstract, so it is fun to be able to illustrate a nature scene while also improving the design of an information graphic. Our figure showing plastics in the marine environment originally came in as a box diagram showing how plastics move through marine environments (2.8a). The boxes indicated reservoirs of plastic debris; the larger blue arrows, fluxes into and out of the marine environment, including potential biodegradation of plastics; and the smaller gray arrows, potential pathways between reservoirs. Although that got the basic information across, I thought it could be improved with a specific setting that showed the different parts of the marine environment and some creatures affected by plastic pollution. However, before I started illustrating, I needed to work out the best way to present the information so the point of the figure would not get lost (2.8b).

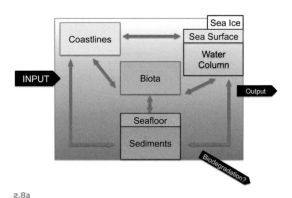

2.8a

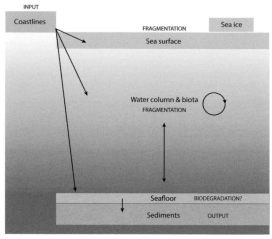

2.8b

I spoke with the author of the article, and she emphasized that the box concept was important to the figure. So, I kept the various elements of the figure in boxes and tried to create a scene they would all fit into (2.8c). At first, I tried a cross section of the ocean, running from the coastline out toward deeper water, and sent it to the author for her thoughts. She refined the way the boxes needed to relate to each other, and I added a bit more of the scene to my second draft.

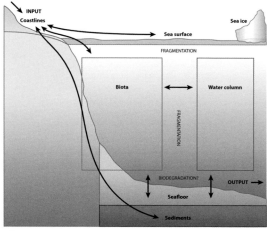

2.8c

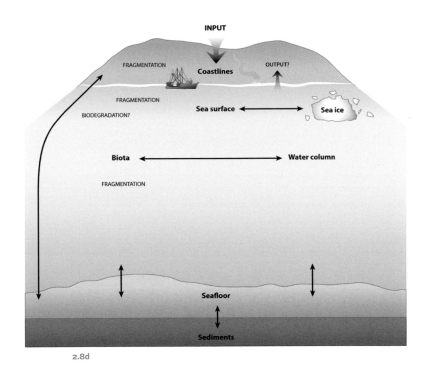

INPUT

FRAGMENTATION **Coastlines** OUTPUT?

FRAGMENTATION

BIODEGRADATION? **Sea surface** ⟷ **Sea ice**

Biota ⟷ **Water column**

FRAGMENTATION

Seafloor

Sediments

2.8d

However, once the second draft took shape, I realized that the view might not work well from a design perspective. I had a lot of empty space in the lower left corner, and I had crowding in a very small amount of space in the design for the coastline and the sea surface. I decided that I should try a different view (2.8d).

In the third draft, I created a view of the environment that would show more coastline and sea surface with less unused space (2.8e). I ran this by the author, and we both agreed that it was the optimal view for the figure. After a little more refinement of the boxes, I could finally start illustrating my scene. I just needed to figure out which location made sense for all the elements that needed to be included, like the coastline, a fishing boat, and ice.

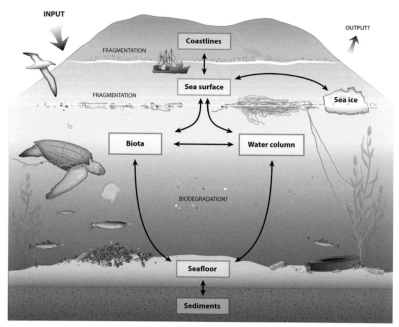

INPUT

OUTPUT?

FRAGMENTATION

Coastlines

FRAGMENTATION

Sea surface

Sea ice

Biota

Water column

BIODEGRADATION?

Seafloor

Sediments

2.8e

We settled on southern Alaska, so I made sure that the animals I drew could conceivably appear in that location. I also included bits of plastic throughout the scene to show that the fragmentation of plastics caused by weathering and biological processes can occur in all reservoirs, especially when exposed to sunlight at the sea surface and along coastlines. In the end, I think it conveys the information in a more effective and compelling way than the original, and the author was very happy with it. It remains one of my favorite *Annual Reviews* figures because I have a marine science background, and the subject matter is close to my heart.

Conceptual

We have just seen how to take physical structure and distill it to its essential pieces. In the examples that follow, we turn this on its head—a bit like Jane's dragonflies! In the graphics that follow, creators take an abstract concept, for example, the ethics of artificial intelligence, and decide how to give it structure.

CHRISTINE DANILOFF, MASSACHUSETTS
INSTITUTE OF TECHNOLOGY NEWS OFFICE
Over the years, I have been lucky to work with Christine, creative director for the news office. She not only comes up with some remarkable designs in ridiculously tight time frames; she also produces videos for the public, explaining research and profiling faculty and students.

The assignment: create a single image for the MIT home page to run with the *MIT News* story "Ethics in Artificial Intelligence." The article would focus on inclusivity in the growing field of AI and the ethical and societal implications of new technologies. This is a subject (and concern) that has been much in the news, and many minds across several disciplines at MIT are doing work around this.

The challenge was figuring out how to marry the themes of ethics and AI in a single image that felt fresh and visually compelling, and that could be done on a tight deadline—I had fewer than five days.

The idea was to connect the human element with technology—a feeling of "many hands at the table" in the development of a new technology. I began with a simple circuit board background and brought in hands joined together as part of the circuitry (2.9a). The themes I chose to focus on to communicate ethics were connectedness, helping, caring for others, and human connection. I felt that incorporating hands into the circuitry, encircling the computer chip—the key to the computer's functioning—was a way to visually suggest these themes in the context of computing.

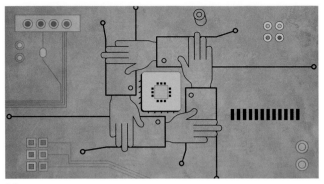

2.9a

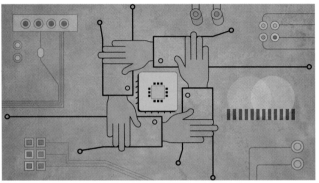

2.9b

It was also a way to convey the idea that human beings hold the power to ensure responsible AI methods and use.

I added the blue gloves thinking about scientists working in a lab who often wear protective gloves (2.9b). I discussed this early composition with the editorial team, and we decided the overall feeling on this was "too male." I also felt that blue gloves, though worn in a lab setting, don't really relate to computer science. The lab coat sleeves might be seen as suits, and overall the hands were too similar and could be more diverse to represent the multiple voices that would need to be involved in the development of AI fields.

I added different colors to the hands but decided the sleeves

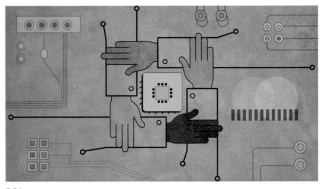

2.9c

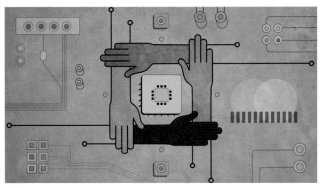

2.9d

still looked too traditionally corporate and could be interpreted as primarily male (2.9c).

In the final image, I removed the sleeves altogether and extended the colors of the hands so that the chip enclosure was hand upon hand, color upon color (2.9d). I think this subtle shift in extending the color created a warmer, friendlier feel to the hands. By removing the subtle details that could be interpreted as gender specific, they became generalized icons of a human touch. In the soft, faded colors of the circuit board, and in the use of washes of color throughout, I tried to create a sense of something that was touched by a human hand rather than an image created by a computer. I feel the image successfully married the two themes of technology and human interaction, and I was happy with the result.

Nik is a truly inspirational artist who makes magic in his cover designs. You previously saw one of my own photographic submissions for a cover that he brought to life. Here he describes a most interesting set of iterations for a concept.

The main challenge when coming up with a cover concept for our vaccines special issue in October 2020 was to celebrate the global scientific effort to develop vaccines against COVID-19 while avoiding the kind of coronavirus imagery that had saturated the media. We wanted to steer clear of syringes and globes and—if possible—not make the virus the main element.

When deciding on a style for the illustration, I wanted to use a theme that conveyed the rapid, bold, inventive work of the scientific community to design, produce, and test vaccines so quickly. Bauhaus was a German modern art movement in the early twentieth century that encompassed all facets of design. It involved radical thinking about how to incorporate core artistic principles into everything from architecture to typography. It promoted problem solving and collaboration between disciplines to work toward functional, effective design solutions. I thought this would work as a perfect theme to represent the endeavor of researchers as they faced the challenges of the COVID-19 pandemic.

I chose the notebooks of Paul Klee—a champion of the movement—as a basis for my cover. I was drawn to Klee's scrawled notes, annotations, and bold color studies. They conveyed experimentation, novel thinking, and rapid work. In particular, I felt that his primary color wheel sketch could be a good starting point (2.10a). It already resembled a virus, with protruding spike proteins!

I also found Klee's bold geometric color study visually striking and noticed how the angles could be used to depict an antibody. I liked the columns of figures, which to me conveyed a sense of process and classification (2.10b).

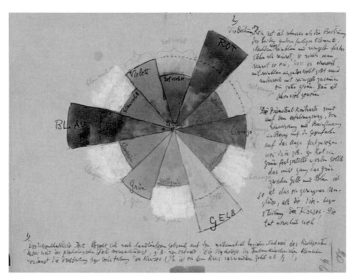

2.10a

2.10b

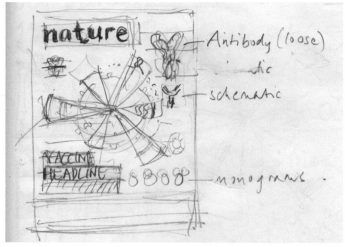

2.10c

2.10d

My early ideas used one main color element, representing the coronavirus with a circular shape inspired by Klee's color wheel (2.10c). I considered surrounding this with pen and ink sketches about viral structure and vaccine types, as well as a mixture of diagrams about antibody form and function (2.10d, 2.10e).

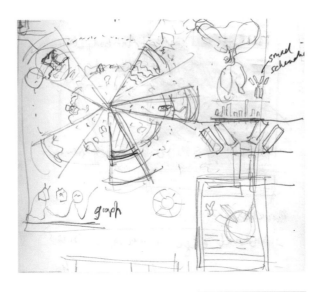

The international journal of science / 22 October 2020

nature

outlook
Antimicrobial
resistance

COVER LINE

The quest for a vaccine to
combat COVID-19

Cover line
Cover blurb goes in
here on three long
lines of loveliness

Cover line
Cover blurb goes in
here on three long
lines of loveliness

Cover line
Cover blurb goes in
here on three long
lines of loveliness

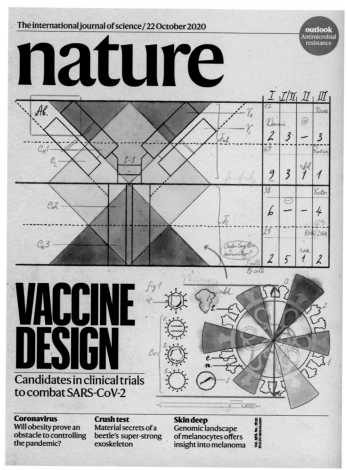

2.10g

In later versions, I used a geometric antibody to replace the color wheel as the main focus of the illustration (2.10f). In these, the virus was more detailed, inked in black and white.

Eventually, we decided to have two main color elements, with a simpler virus used as the starting point of three parts (2.10g). The first part would use the wheel to convey studies of the virus, with wedges to represent how its structure and genome have been rapidly resolved and analyzed. To the side of these wedges, we depicted potential vaccine candidates using an infographic I had produced in April as a basis for these elements.

The second part was the antibody representing immune response and the goal of immunity. The third and final part was scrawled columns of figures to the right of the antibody figure, depicting the numbers of vaccines at various stages of clinical trials (indicated by roman numerals). I then added notes and annotations in pencil to give a sense of ongoing study and rapid progression through the experimental, testing, and clinical trial processes. The result is an imagined page from a notebook of vaccine design. We hoped this cover served as a shot in the arm [giggle—*Felice*] to those weary of seeing conventional depictions of the virus and inspired hope and creativity in the research community and beyond.

CATHERINE ZUCKER, WITH ALYSSA
GOODMAN, HARVARD UNIVERSITY

I met Catherine Zucker when she was a PhD candidate at Harvard University's Center for Astrophysics, with Alyssa Goodman, a professor of applied astronomy and a dear friend and colleague, as her adviser. Since then, Catherine has become a formidable astrophysicist on her own. In this case, Catherine is communicating research on interstellar filaments, which are precursors for star formations. I considered categorizing this example as structural, but I'd argue Catherine is representing processes—the formation of these filaments. Thinking about your goal—to communicate structure, concept, or data—is helpful for understanding your science.

The ultimate goal of this graphic was to illustrate similarities and differences among recently discovered interstellar filaments, which stretch tens of light-years and are a precursor for star formation, in our Milky Way galaxy. Because our solar system is embedded inside the Milky Way, it is incredibly difficult to map out our galaxy's structure from within. These filaments may hold the key to this mapping, as they may form in "special" places in our Milky Way (along its characteristic spiral pinwheel pattern) and thus be used as key landmarks in constraining its structure.

It is important to understand the physical properties of these filaments to understand where and how they form in our Milky Way galaxy.

The labels ("MW Bone," "LS Herschel," "MST Bone," and "GMF") represent the filaments described by four independent published studies. Each of the four studies used different selection criteria and methodology to identify the filaments. We sought to answer the question, Are all four classes of filaments the same? That is, did they form in the same way?

The colors themselves have no relation to the physical structure of the filaments, but we preserved the color scheme throughout the paper to allow readers to differentiate the various classes easily.

The first version of the graphic showed only the overall shape and size of the filaments, with no contextual information about where they lie in the galaxy (2.11a). Each rectangle represents the average length and width of the filaments described in one of the four studies (approximately ten to twenty filaments each), such that their aspect ratio is preserved. However, the rectangles give no information about the variation in length and width within each sample.

The text labels under the skinny rectangles indicate the class name, the average aspect ratio (e.g., 34:1), and the average length (e.g., 40 parsecs, or pc) of filaments in the class. A parsec is equivalent to three light-years.

MW Bone: 34:1, 40 pc

LS Herschel: 34:1, 57 pc

MST Bone: 11:1, 30 pc

GMF: 8:1, 120 pc

50 pt = 10 pc

2.11a

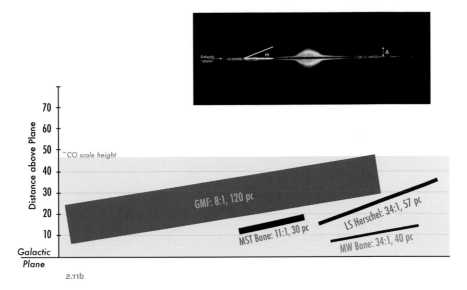

2.11b

The text at the lower right shows the scaling between pixel units in Keynote (the Apple equivalent of PowerPoint) and actual physical units in the galaxy: fifty pixels in Keynote is equal to ten parsecs in the galaxy, or about thirty light-years.

In the next iteration, we again included the average length and width for each class of filaments and preserved the aspect ratio, but we added information on where the filaments are situated in the larger context of the Milky Way (2.11b). The orientation of each rectangle indicates how the filaments are oriented with respect to the "plane" of the Milky Way's disk. This is equivalent to measuring the angle θ in the schematic at the upper right.

The vertical axis of the graph shows the distance between the plane and the filament, which I have marked with the symbol Δ. The displacement is not to scale—all the filaments are actually inside that very thin disk, and we would have to zoom in very far to see them.

Where the filaments are placed along the x axis (the "galactic plane") has no spatial meaning. We care only about their morphology, placement along the y axis, and angle relative to the x axis. The gray box in the background labeled "CO scale height" is the typical thickness of the carbon monoxide gaseous disk that is used as a tracer of the interstellar gas from which the filaments form. This gives a sense of how displaced the filaments are with respect to the general disk of material constituting the Milky Way galaxy.

In the final version, we depicted every filament as its own rectangle with its actual measured length, width, displacement from the galactic plane, and orientation with respect to the galactic plane (2.11c). The filaments end up looking a lot like matchsticks, so we informally refer to it as the "matchstick diagram." More important, this allows us to see the variation in filament properties within a given class.

This version provides more detail on where each individual filament lies with respect to the galactic plane. Filaments with negative y values lie "below" the galactic plane, and filaments with positive y values lie "above" the galactic plane. We also removed the gray box from the previous slide showing the typical thickness of the plane, mostly for aesthetic reasons but also because most astronomers tend to have an idea of how thick the general plane of the galaxy is. As before, the x axis has no physical meaning, but here we arbitrarily divided it into four panels and "dropped" the filaments into each panel at random x positions.

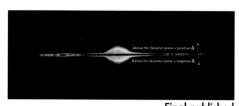

Final published version:

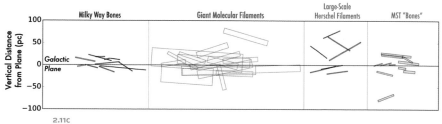

2.11c

Viewing the filaments in this way, we can see that the skinniest filaments that are closest to the plane and with the smallest inclinations with respect to the plane are the "Milky Way Bones" class of filaments. Other filaments tend to be thicker and farther from the plane. This suggests that the filaments formed via different physical mechanisms.

Numerical

In this final section, we see just a few examples of how talented researchers and designers think about representing large data sets. I can provide only a brief introduction to the concept of data visualization here—it is a subject worthy of its own book. I want you to understand that there is a difference between visualizing data and using a data visualization to communicate. It's one thing to see a ton of numbers spewed out in a data set. It's quite another to find a way to make sense of it all.

MARTIN WATTENBERG AND FERNANDA
VIÉGAS, HARVARD UNIVERSITY
Martin and Fernanda are both professors of computer science at Harvard University School of Engineering and Applied Sciences. Their work together is truly remarkable, bringing a much-needed aesthetic to a discipline that desperately needs a makeover. I have been following their work for years.

Inspired partly by a blustery New England winter, the two of us set out to visualize the wind. It was the kind of challenge we love: taking something you can't see and giving it visual form. Although there's a long history of charting the wind, from seventeenth-century nautical charts to the labs of present-day scientists, we were searching for something new, a more dynamic way to convey the mystery and power of an elemental force.

The design process we describe took a couple of months and presented us with a number of dead ends. We nearly gave up, thinking there was nothing new we could add to the historical arc of wind mapping.

When we begin any visualization, our first concern is the data, in this case, surface wind data from the National Digital Forecast Database that shows surface wind speeds (2.12a). We made the first graphic with one goal in mind: to make sure we'd parsed the wind data correctly. Each line shows the direction and magnitude of the wind at that point. The green-on-black color scheme is our way of signaling (even just to ourselves!) that this is a draft.

Our first visualization was a set of dots moving according to the wind direction (2.12b). Intuitively, it made sense to us to try to depict wind as a set of particles moving on the screen. However, the resulting visualization was unreadable. More than anything else, it looked like dust blowing around on the monitor.

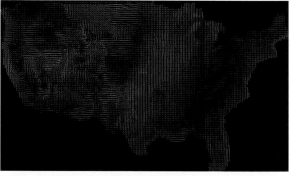

2.12a

2.12b

So we tried something completely different: we drew a colorful grid and distorted it according to the wind data (2.12c). The result looked like a plastic map that had been put in the oven for a few hours. On this day, a strong northerly flow stretched the map in that direction.

We gave up on the melting colors and tried a stark black-on-white view, drawing black-and-white trails that followed the wind (2.12d). This view, which was reminiscent of a zebra-skin rug, seemed to hold promise—but it was static, and we missed the animation. At a conceptual level, a key flaw was that you

2.12c

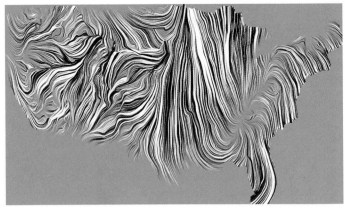

2.12d

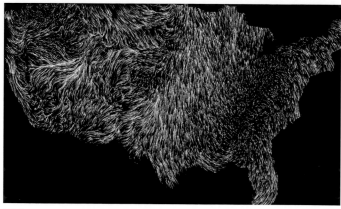

2.12e

wind map

July 21, 2021
2:19 pm EST
(time of forecast download)

top speed: **26.9 mph**
average: **7.6 mph**

1 mph

3 mph

5 mph

10 mph

15 mph

30 mph

2.12f

couldn't tell which direction the wind was actually going: an east wind and a west wind would look identical. At this point, we felt stuck—but after several more conversations, we pressed on.

We went back to an animation, this time leaving trails behind dots (2.12e). As you look at the image, imagine it in motion, with lots of dots in the Midwest surging to the north. We were getting close!

The final image is from the Wind Map website (hint.fm /wind), which has been running since 2012 and is continuously updated to show current conditions (2.12f). Since then, we've created many variations ranging from print work to museum installations.

Because we approached this project primarily as artists, some aspects of the design may seem unusual. The map is grayscale, leaving a major visual dimension, color, completely unused. Our goal was to help viewers focus on the patterns of animation over everything else. We did not want boundaries to interrupt the flow of the animation, so we omitted state borders and instead used a handful of cities to orient viewers.

The wind map seemed to resonate with many people and took on a life we never expected. Others have extended this idea in beautiful ways, including interactive 3D globes and incorporating it into television weather broadcasts. People have written to tell us they've shown the map in classes and used it to understand the migration of butterflies. And on blustery New England days, we still often look at it ourselves.

ADOLFO ARRANZ, CREATIVE DIRECTOR OF
THE *SOUTH CHINA MORNING POST*
Visualizing big data for print is a tricky business. As you navigate through Adolfo's iterations, you will see the challenges he faced. I think his final graphic works beautifully because of its simplicity and success in immediately giving us a sense of the enormousness of the numbers. What makes it even more interesting is that the final visual is more compelling in print than it would have been if it were viewed on a screen.

In January 2018, the Aviation Safety Network (ASN) released a remarkable report. The previous year had seen only ten fatal commercial airline accidents, with the loss of forty-four lives. It was a new record for air safety. Just ten years earlier, there were almost eight hundred fatalities in commercial plane crashes.

As soon as I saw the news, I proposed making a graphic that would fill the back page of the newspaper where I work, the *South China Morning Post*. The idea would be to publish it in the shortest possible time. The ASN site broke down the data by type of accident, date, the phase of the flight in which the accidents

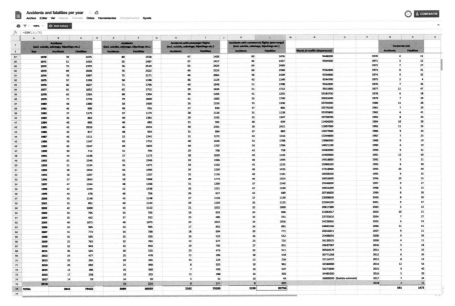

2.13a

occurred, and other interesting parameters (2.13a). But two figures were the basis of the news: in 2017, there were 36.8 million
flights and only ten fatal accidents. So I focused on those two
figures to make something simple and visually powerful. Instead
of showing the typical statistical graphs, as ASN had done on the
website, I tried to imagine how I could show that data in an easy-
to-understand way. The first problem was figuring out how to
compare two such disparate figures—visualizing 36.8 million and
10 in the same graphic is very complicated.

I started with the number of flights. Because I wanted this to
look distinctive, I tried using points or countable objects on the
page (2.13b). I'm lucky that the *Post* has a rather large page (32.5
by 53 centimeters). To show those 36.8 million points, I asked my
colleague Pablo Robles if he could generate a random pattern
with that many black dots. And he did.

The first attempt gave an undesirable result: the points were
too concentrated and could not be differentiated, creating an
utterly black page.

Next, I tried dividing the number of points by 10, 100, and

A escala 100 personas por punto:

2.13b 2.13c

200, with disparate results. Consider two attempts at reducing the dot density: in the first, each point represents ten people, and in the second, each dot represents one hundred people (2.13c).

The image with a 1:100 representation, or 368,000 dots, seemed the most suitable for displaying the data (2.13d). The hundreds of thousands of points are differentiable and do not stack, and the page manages to be more compact and exciting. So I decided to use that ratio. The important thing was to put into perspective the quantitative dimension of that information and on a full-size page.

Now it was time to compare those ten fatal accidents with all those points. With each dot representing one hundred departures, the ten accidents in 2017 would account for just one-tenth of a single dot. So I used zoom mode to enlarge one dot and divide it into one hundred dots, each representing a single flight. I highlighted the fatal crashes in red (2.13e).

In another draft, a text box and too many callouts cluttered the design (2.13f).

Each dot on this page represents 100 departures. Last year's accidents account for one tenth of a single dot

Enlarged dot

Total accidents *with fatalities*
10

79 fatalities in 2017 [2]

2.13f   2.13g

For the final version, I moved the headline and introductory text to the top and removed as many other elements as possible to maximize the visual impact (2.13g).

This page was produced in a day and a half with help from my colleagues Pablo Robles, Marcelo Duhalde, and Darren Long. It won a medal at the Society for News Design Annual Awards and various other international awards.

This type of design, in my opinion, would be challenging to publish in many newspapers. Many editors would reject using an entire page to show 368,000 dots, but the *South China Morning Post* is open to such proposals.

Slotting visuals into just one category might not be helpful when starting out with one of your own visuals. However, presenting these examples in this way might encourage you to take your time and consider one or two of these categories for your own work.

3

Graphic Submissions

Figuring the Figures

Any figure you create will, by necessity, be only a representation of something, not the thing itself. As you've just seen, to represent your science visually, you have to make decisions about what to include and what to leave out—to abstract the concept down to its essence and then give it a physical representation in the form of a graphic. You also have to make careful decisions to ensure both clear communication and scientific integrity.

After years of running workshops and working with researchers, I have concluded there is one overarching principle that continues to challenge scientists when creating figures to submit to a journal—to *simplify* the graphic.

Before making a figure, try to clarify in your own mind what it is that you want to show. What do you want me to see? Is it evidence for a relationship in your data? Details of a structure? A process that happens over time? A comparison of some kind? All of the above? (That last one was a trick question: if you are trying to show all of the above, you are trying to show too much.) Keep the following in mind as you start working on your figure and continue to reevaluate its clarity.

- Are you including data just because you have it?
- Do all the elements of your graphic support your story?
- Is there redundancy? Must you include a particular image or graph that you plan to show elsewhere in another form?
- Can some of the data go into the supplemental materials?

CAN YOU EDIT OUT GRAPHICAL DISTRACTIONS?

- Unless the journal requires parentheses in your figure label, there is no need for them.
- Do you really need *all* those tick marks on the graph's axes, or is that a default setting in your software?

Understand that space is your friend: the more blank space in your figure, the easier it will be for viewers to see the components.

There is more to design, however, than just deciding what to leave out and creating space. This chapter addresses five additional strategies that can increase the clarity of your graphics:

Color: deciding where to add color to your figure and when too much color is confusing

Composition: organizing the figure's pieces to create a visual hierarchy of information—that is, where you want the viewer to look first

Layering: overlaying components to visually add information and facilitate comparisons between data sets

Labeling: adding text that adds clarity without adding clutter

Refining: fine-tuning your figure with small changes that address one or more of the above

An important note when considering all this advice is that journals have their own rules limiting layout and font choices. You'll have to look up the requirements for the specific journal you're submitting your work to. They are surprisingly different. Here, you will mostly see iterations of graphical ideas and how simple changes can clarify the communication of particular information.

This chapter includes roughly forty before-and-after depictions, most of which were of researcher drafts created for my workshops. The first image in each pair is a figure created for our group discussions. The second figure incorporates suggestions from me or from workshop conversations. You'll also see a few "generic" examples created for the chapter.

For each pair, start by taking a good look at the original. Pause to think of your own suggestion on how to clarify the graphic. Then take a look at the revision to decide whether you agree with the changes. You might not! Again the goal is to show you which elements you might change and how they change the design. The "best" changes might depend on the audience—there are few hard-and-fast rules.

We'll start with color. Why do I begin with color? It is one of the most challenging aspects of creating graphics for scientists. It is where I see the most errors impeding clear communication.

Color

Using color is universal in all disciplines and at various scales. Consider a visualization of methane on Mars in the summer in which orange and red show higher concentrations (3.1).

And consider a map of the cerebral cortex, the outer mantle of the brain, where researchers have discovered 180 distinct areas in each hemisphere (3.2). In this case, the colors represent distinct areas of the brain.

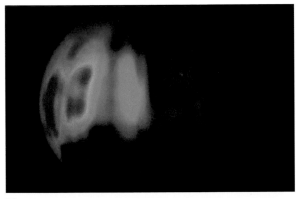

3.1

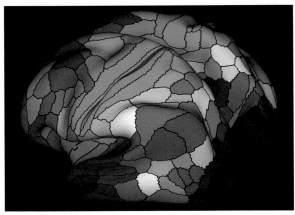

3.2

Both of these depictions lean toward a rainbow color palette, which has often been the case. But that's not necessarily the best choice. It's not that it doesn't work to differentiate certain areas, because both of these do both. The issue, for me, is an aesthetic one. I am not drawn to this palette because it is one that I have seen for too many years. Happily, we are seeing more interesting and communicative combinations these days, which I encourage you to consider.

COLOR CHOICES

Often, junior researchers will ask which colors they should use. Frankly, I have a hard time responding. If you do a search, you will find an overwhelming number of articles on color perception and which choices produce which perceptual results. For example, can the human eye distinguish the selected colors? Is the color palette accessible for color-blind readers? The websites Better Figures and Color Brewer are definitely worth a look, but they can be overwhelming, and I am not convinced that the choices made should be based on perception alone, as these sites recommend.

An approach those websites suggest, which I agree with, is using the same color family at various saturations when your data is about showing differing quantities of the same thing. For example, the mechanical engineer and data visualization researcher Haluk Akay selected four similar colors to show a model of infectious spread (3.3). They work well, and we can easily read the changes over time.

Consider a similar example (3.4). The colors in a calendar published by the *New York Times* and continually updated online to show the rise and fall of COVID-19 cases *became* intuitive to that publication's readers over time. At first, seeing only one or two months of data, the color choices were not obvious. But after several months of more data coming in, the color choices made more sense. They are all from same color family, and we immediately understand that during certain months, cases increased drastically.

Graphical modeling of infectious spread

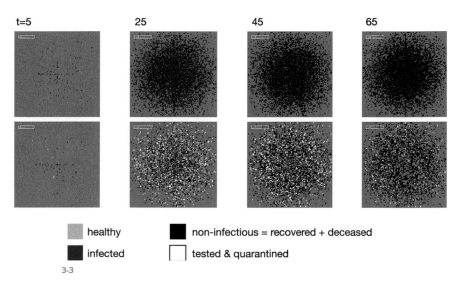

t=5   25   45   65

healthy

infected

non-infectious = recovered + deceased

tested & quarantined

3.3

Average cases per capita in Middlesex County

Fewer   More

3.4

However, one component rarely discussed with regard to choosing color is personal aesthetics, which I am convinced plays an important role. We bring our personal histories with us when we make decisions about color. Experiences we had as children, such as the magazines we grew up with or the crayon colors we gravitated toward in the box, left impressions on us that influence us later in life. If I suggest to a researcher that combining fuchsia and chartreuse doesn't work—my professional opinion, including that most people would find it garish—it still might not be enough to change her mind if she has a strong personal attachment to those colors. We need to realize our personal color biases because they may lead us to color choices that don't work well for others. My advice is to take a look at how various publications use color to get a sense of those publications' aesthetic. Moreover, look at nonscientific publications and judge for yourself how well their uses of color work.

In a map from the *Economist*, I would not have thought to choose beige and green, but I find the color choice appealing (3.5). The colors are also easy to read and intuitive. Note how they connect to the meaning of the data, with green corresponding with high soil fertility. Color helps make this complicated map easy to grasp: it shows where American bombing fifty years ago still shapes Cambodian agriculture, because farmers avoid fertile areas with soft earth that may contain unexploded bombs.

When I saw a map in the *New York Times*, I paid attention because of the subject matter and also because I liked the choice of colors purely for aesthetic reasons (3.6). Frankly, I think the colors are beautiful together. The graphic visually describes how the Minneapolis police used force against the Black community at a rate seven times that against whites. Larger circles indicate that more incidents occurred.

Cambodia, soil fertility in 2008 and American bombs dropped 1965-73

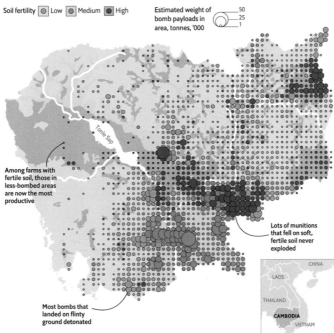

Soil fertility ● Low ● Medium ● High

Estimated weight of bomb payloads in area, tonnes, '000

Tonle Sap

Among farms with fertile soil, those in less-bombed areas are now the most productive

Lots of munitions that fell on soft, fertile soil never exploded

Most bombs that landed on flinty ground detonated

CHINA
LAOS
THAILAND
CAMBODIA
VIETNAM

3.5

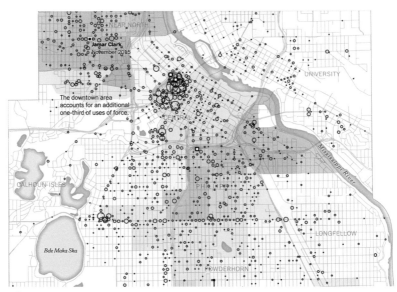

NEAR NORTH

Jamar Clark
November 2015

UNIVERSITY

The downtown area accounts for an additional one-third of uses of force.

CENTRAL

Mississippi River

CALHOUN ISLES

PHILLIPS

LONGFELLOW

Bde Maka Ska

POWDERHORN

3.6

Color choices that are somewhat intuitive provide the viewer with a handle on the graphic. In a *Cook Political Report* visual, we clearly understand the political color choices of blue and red and gray (3.7). Placing the blue circles on the left reinforces the graph's political nature. There's a description at the top, but the intuitive colors make it easy to understand even without it. One note to consider is that this visual is intended for an American readership. Other parts of the world (including most of Europe) use red to represent the left and blue to represent the right. This would be a confusing graph, for example, if it represented UK politics.

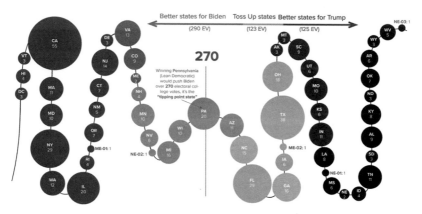

This chart was inspired by 538's 2016 Winding Path chart and Priorities USA's snake chart.

3·7

A graph from mechanical engineer Gang Chen's lab shows evaporation rates of water under light-emitting diodes (LED) illumination of different wavelengths as a function of the LED intensity (3.8). His use of color matches the actual color of those wavelengths.

3.8

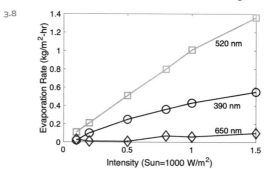

Combining colors to form an in-between color can clarify some types of data. The color choices in a figure can be politically relevant, but the point is the combination of the two, a shrinking purple area of overlap that indicates how the United States has become more polarized over the years based on survey data on political values collected by the Pew Research Center (3.9a, 3.9b).

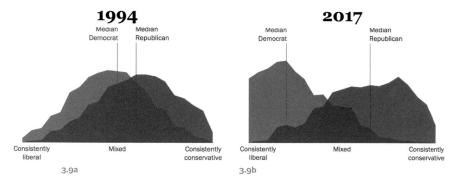

Consider another, more generic example of how color can show the area of overlap between two peaks (3.10). For further clarity, the peaks are outlined.

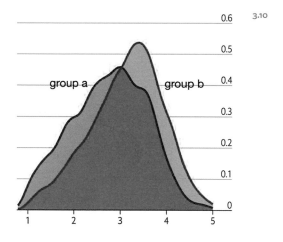

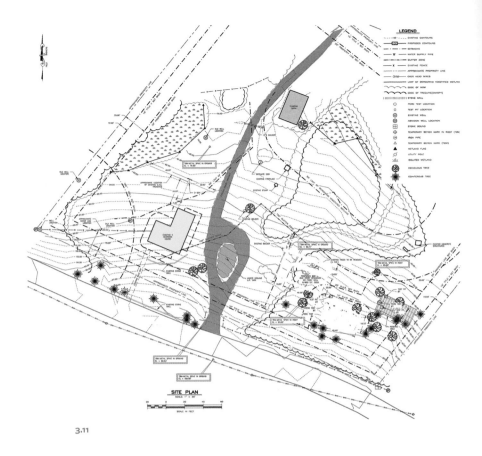

3.11

BRINGING ATTENTION

Perhaps the most obvious use of color is to bring attention to a specific area in a graphic, as we see in this architectural plan of a site (3.11). The three colored areas are described in detail in a proposal. Often, however, researchers use so much color that the reader cannot figure out where to look, as you will see in some of the examples that follow.

In the following examples, original figures are presented first with the revised versions after.

Researcher: Eva Wang

Purpose: To show adhesion forces measured at different temperatures.

Suggestion: Switch the color palette to use intuitive colors: red for higher temperatures and blue for lower temperatures (3.12a, 3.12b).

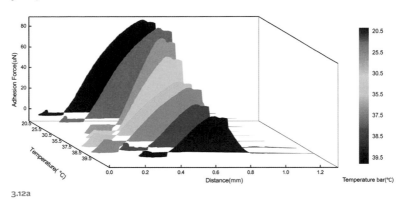

3.12a

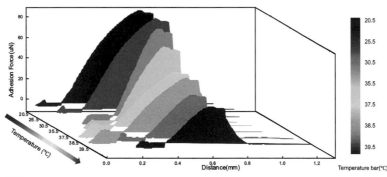

3.12b

Researcher: Klavs Jensen

Purpose: To show how material moving through a device is subject to different pH gradients.

Suggestion: Use intuitive pH colors: red for acid, blue for basic, mimicking traditional litmus paper colors (3.13a). Also, insert arrows to show the direction of flow (3.13b).

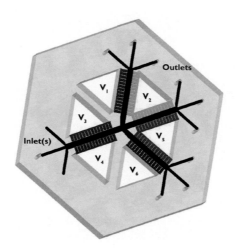

3.13a

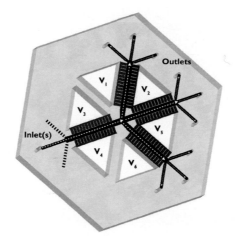

3.13b

Researcher: Anthony Barberio

Purpose: To show a process for coating liposomes to improve drug delivery.

Suggestion: Minimize color to bring attention to the coating material (3.14a, 3.14b).

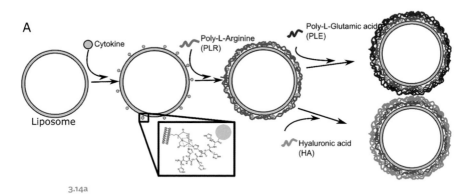

3.14a

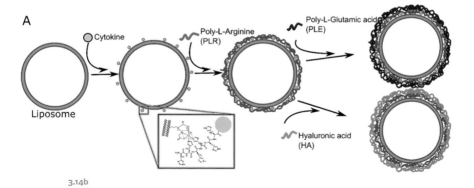

3.14b

Researchers: Students in a Sports Mechanics class taught by Anette Hosoi and Christina Chase

Purpose: To show shots by the US Women's soccer team in the World Cup final.

Suggestion: Make the background grayscale for better visibility of data points (3.15a, 3.15b).

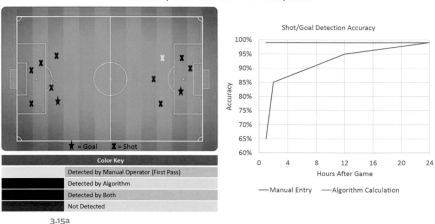

3.15a

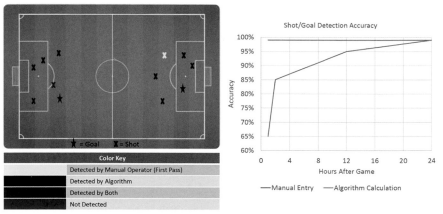

3.15b

Researcher: Heeding Roh

Purpose: To show a "grafting" modification to polyurethane.

Suggestion: Make the background structure grayscale and delete color from the labels to focus attention on the modification (3.16a, 3.16b).

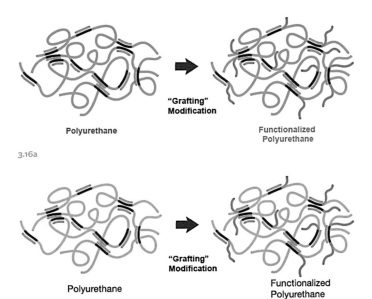

"Grafting"
Modification

Polyurethane

Functionalized
Polyurethane

3.16a

"Grafting"
Modification

Polyurethane

Functionalized
Polyurethane

3.16b

Researcher: Aristotle Grosz

Purpose: To show data along three axes.

Suggestion: Delete the distracting colors in the axis labels to emphasize the colored graphs (3.17a, 3.17b).

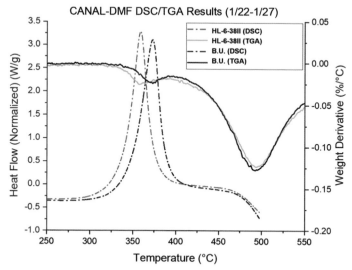

3.17a

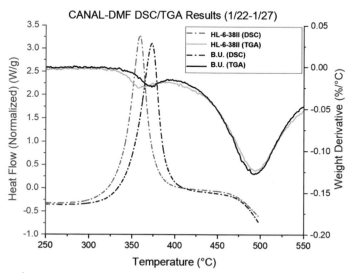

3.17b

Researcher: Sara Wu

Purpose: To show a surgical process, highlighting the role of a collagen sponge.

Suggestion: Delete any unnecessary color to highlight sponge (3.18a, 3.18b).

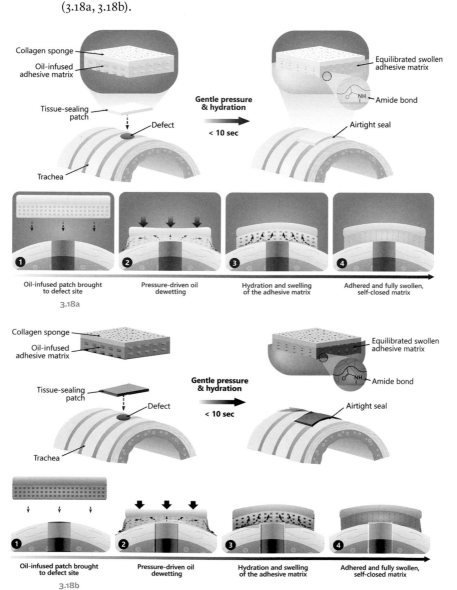

Generic example

Purpose: To make comparisons among related data.

Suggestion: Delete unnecessary color. Consider a simple gray-scale, especially if using this chart with additional colored components in a figure (3.19a, 3.19b).

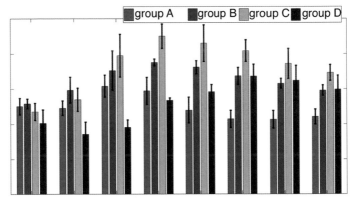

x axis, increasing value

3.19a

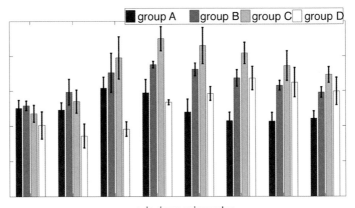

x axis, increasing value

3.19b

Researchers: R. Cant et al.

Purpose: To compare specific areas of an explosion, highlighting two variables: the top represents cellular structure in terms of pressure, and the bottom shows the reaction progress.

Suggestion: Use color only to highlight the relevant parts (3.20a, 3.20b).

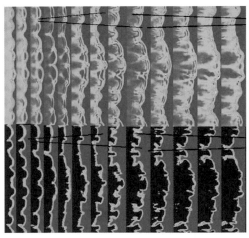

3.20a

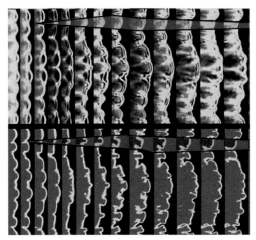

3.20b

Researchers: D. Eigler et al.

Purpose: To show the placement of atoms in a corral and the quantum effects within the corral.

Suggestion: Make the image grayscale so that the atoms and the topological elements have equal visual weight; this communicates to readers that all components have the same importance (3.21a, 3.21b).

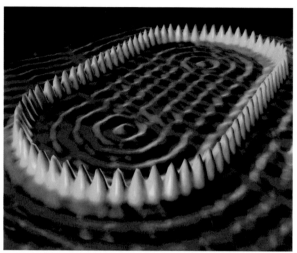

3.21a

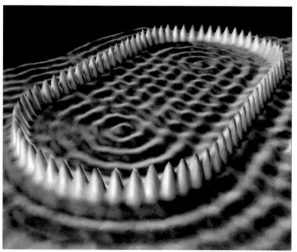

3.21b

Researcher: Jordan Sosa

Purpose: To describe various charged areas in the porous electrode theory.

Suggestion: Do not repeat colors with unrelated items to prevent confusion. For example, note the red square in the first image, which doesn't relate to the red reactant circles. The same mismatch exists for the blue arrows and square and the blue product (3.22a, 3.22b).

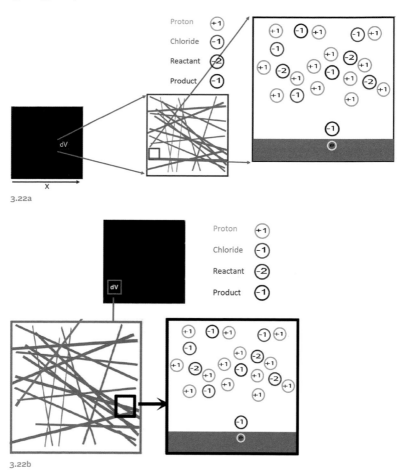

3.22a

3.22b

Researchers: Tonio Buonassisi and Yun Seog Lee

Purpose: To bring attention to the nanometer-thin buffer layer between a cuprous oxide solar cell absorber and the transparent contact layer.

Suggestions: Color only the buffer layer, and delete unnecessary graphical elements such as the parentheses (3.23a, 3.23b).

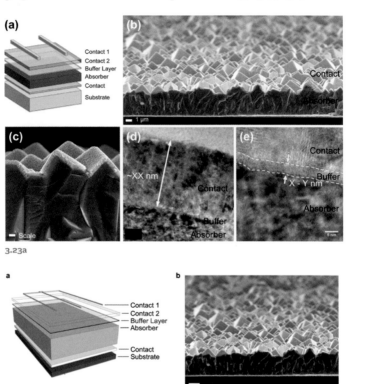

3.23a

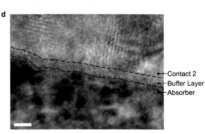

3.23b

Data from the Web

Purpose: To compare annual global carbon dioxide emissions among countries.

Suggestion: Do not repeat colors unless there is a reason to do so, and remove unnecessary frame and lines from the legend (3.24a, 3.24b).

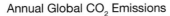

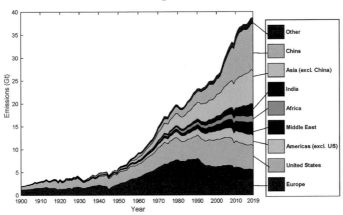

3.24a

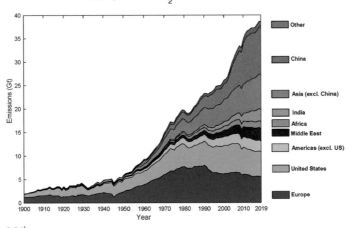

3.24b

Composition

Organizing the components of your figure can be a powerful tool to clarify your message. For starters, note one clever way of placing various materials on a graph (3.25). We immediately see the purpose of Shahrzad Ghaffari Mosanenzadeh's figure: to compare the materials' weights, mechanical strengths, and noise reduction abilities with those of an iron foam and silica aerogel composite.

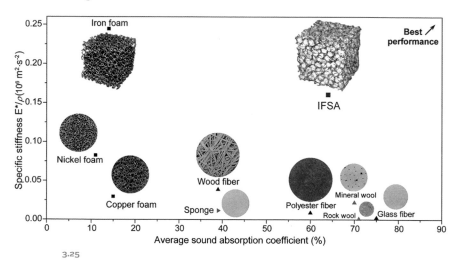

3.25

I'll start with one of my own redos.

Researcher: George Whitesides

Purpose: To show various self-assembled structures for a grant submission.

Suggestions: I arranged their images according to size and added a single scale reference. In addition, I used a schematic to label components and minimize clutter (3.26a, 3.26b).

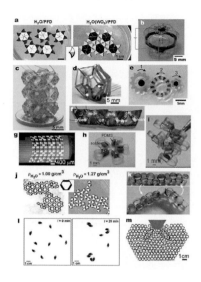

3.26a

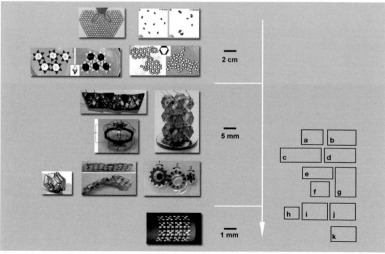

3.26b

Researcher: Yanfei Guan

Purpose: Compare one experimental approach ("this work") with others.

Suggestions: Recompose the titles and descriptions of pros and cons for each approach (3.27a, 3.27b).

A. Chemical meaningful descriptors:

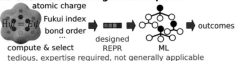

tedious, expertise required, not generally applicable

B. Machine learned molecular representation:

high dependancy on the training set

C. This work:

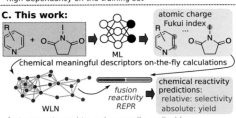

fast, accurate, end-to-end, generally applicable

3.27a

A. Chemical meaningful descriptors	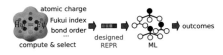	tedious, expertise required, generally not applicable
B. Machine learned molecular representation	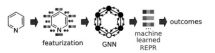	high dependancy on the training set
C. This work	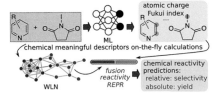	fast, accurate, end-end, generally applicable

3.27b

Researcher: Junsoo Kim

Purpose: To describe how the pit membrane of a tree served as inspiration for the structure and process of a chemical pump.

Suggestions: Reorganize components to be read from left to right (3.28a, 3.28b).

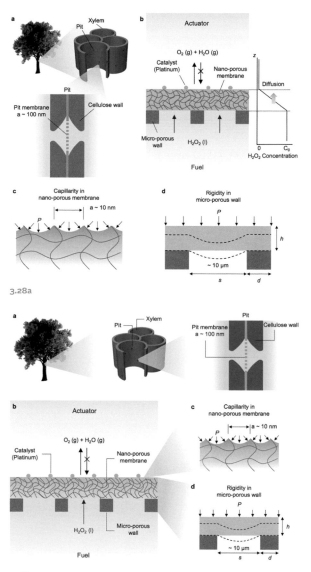

3.28a

3.28b

Researcher: Joy Zeng

Purpose: To show the aims at various steps of a research project.

Suggestions: Reorganize titles and components to make the sequence clearer (3.29a, 3.29b).

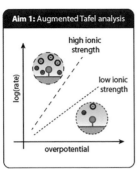
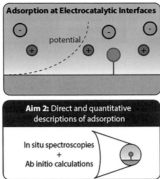
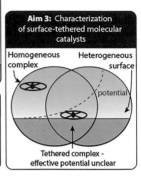

3.29a

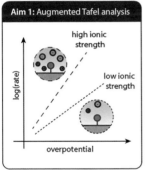
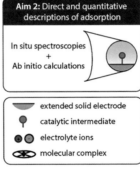
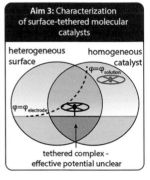

3.29b

Researcher: Kaylon Overholt

Purpose: To show how data will be collected.

Suggestions: Create a visual hierarchy in which data appear more important by enlarging their representation and reducing others (3.30a, 3.30b).

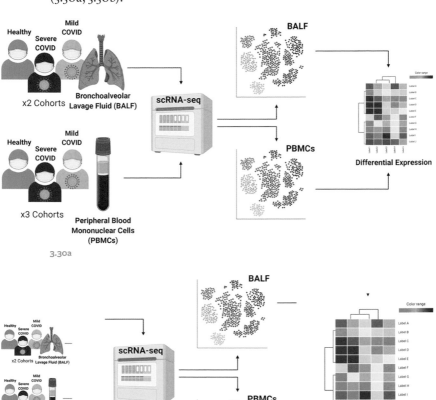

3.30a

3.30b

Researcher: Sarah Shapiro, of the Doyle Lab

Purpose: To show how the signal in a biotin-streptavidin bioassay can be increased by increasing the ratio of the surface area of a hydrogel particle to the two-dimensional projected imaging area and then used for analysis.

Suggestions: Sarah reorganized the components to create a more horizontal layout, making a better fit for the journal. She also changed the colors to emphasize structure and eliminated one of the two graphs (3.31a, 3.31b).

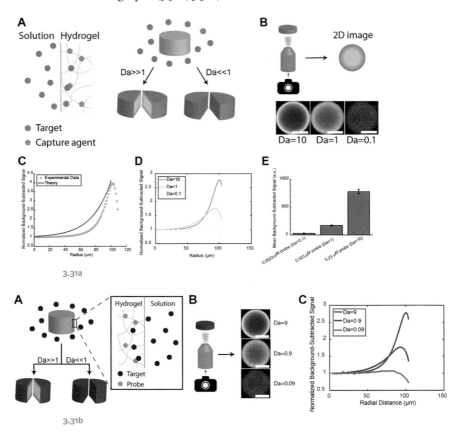

3.31a

3.31b

Layering

Adding an illustration or layering one set of graphics over another can often draw attention to a particular area of a figure or facilitate a comparison between two sets of data.

Here's an overlay I made quite a while ago (3.32). I made the photograph of the apparatus, but including the laser would have been impossible to photograph, so I added an illustration creating a photo illustration.

Anthony McDougal studies structure formation in live butterflies with an innovative technique in imaging. The image he sent outlines the area he wants us to focus on (3.33).

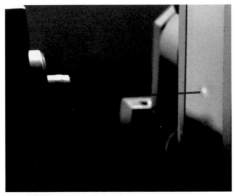

3.32

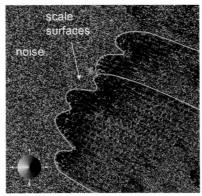

3.33

Researchers: R. Bruckner et al.

Purpose: To compare the regional thinning of brain tissue in normally aging brains with brains affected by Alzheimer's disease.

Suggestions: Recolor the normally aging brain so that when it is overlaid with the Alzheimer's brain, one can more easily make the comparison. Reorder the labels to create more uniformity (3.34a, 3.34b).

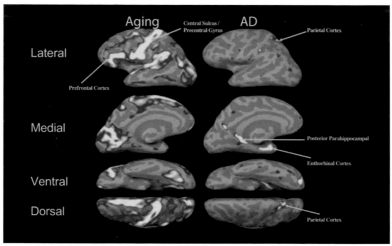

3.34a

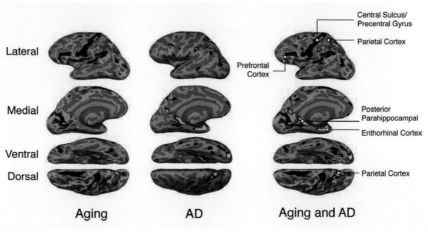

3.34b

Researchers: Jay Pasachoff et al.

Purpose: To show the detailed structure of the Eagle Nebula.

Suggestions: Outline the area in the larger image with a related color to help readers understand where the detail comes from (3.35a, 3.35b).

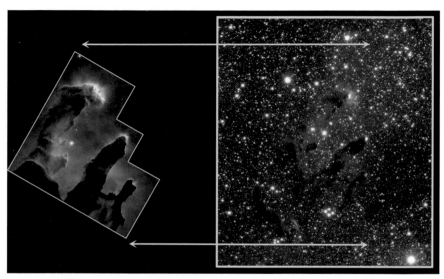

3.35a

3.35b

Researcher: Jongwoo Lee

Purpose: To compare two sets of data.

Suggestions: Layer one set over the other since both graphs share similar axes (3.36a, 3.36b).

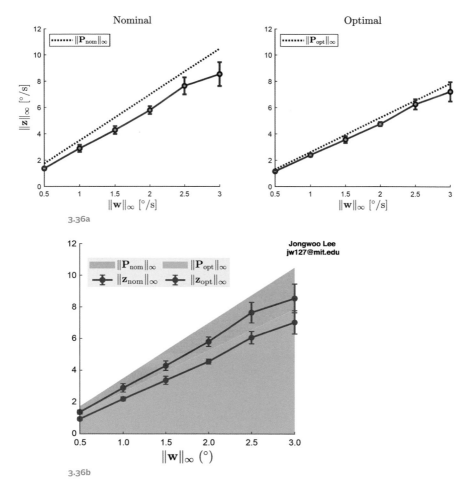

3.36a

3.36b

Jongwoo Lee
jw127@mit.edu

Researcher: Keith Legrand

Purpose: To describe data collection by a camera-equipped aircraft tasked with finding and tracking vehicles on the ground. Using probabilities of where vehicles might be found and where they might travel (top image), the aircraft autonomously points its camera to collect the most informative pictures.

Suggestions: Overlay outlines of collection areas on the terrain in the lower image and match the line style of the outlines with those in the upper image (3.37a, 3.37b).

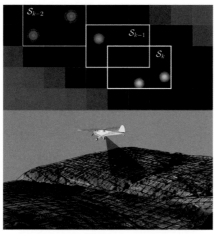

3.37a

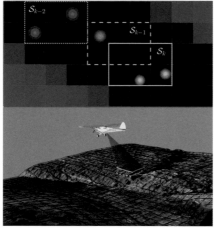

3.37b

Researcher: Lup Wei Chew

Purpose: To compare two sets of data.

Suggestions: Overlay one set on top of the other since both have the same axis (3.38a, 3.38b).

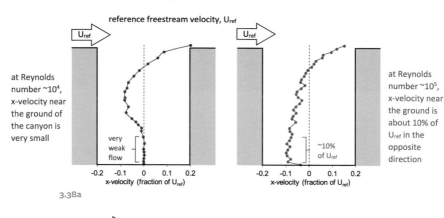

at Reynolds number ~10⁴, x-velocity near the ground of the canyon is very small

at Reynolds number ~10⁵, x-velocity near the ground is about 10% of U_ref in the opposite direction

3.38a

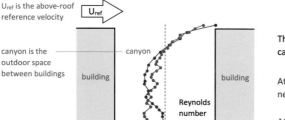

U_{ref} is the above-roof reference velocity

canyon is the outdoor space between buildings

The x-velocity near the ground of the canyon is <u>Reynolds number dependent</u>

At Reynolds number ~10⁴, x-velocity near the ground is almost zero.

At Reynolds number ~10⁵, x-velocity near the ground is about 10% of U_{ref} in the opposite direction.

3.38b

Labeling

In our workshops, I've noticed that labels and legends often seem to get done at the last minute and without much thought. For example, students generally accept the default settings in software without making choices best for their data. Here are a few examples that might get you to think differently.

BEFORE-AND-AFTER EXAMPLES

Generic example

Purpose: For a slide presentation, compare two systems over time.

Suggestion: Move labels closer to the corresponding lines on the graph. Delete unnecessary tick marks on the *x* axis to create more white space (3.39a, 3.39b).

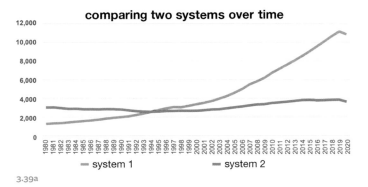

3.39a

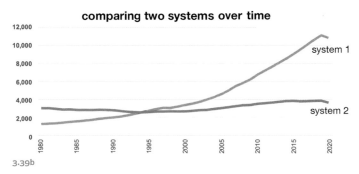

3.39b

Researcher: Connor Coley

Purpose: To label parts of a new apparatus for laboratory automation.

Suggestion: Straighten the label lines to make the labeling more uniform (3.40a, 3.40b).

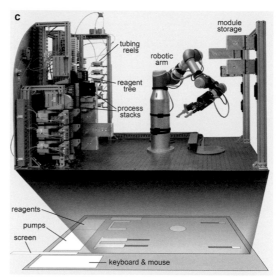

3.40a

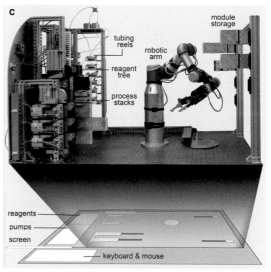

3.40b

Researcher: Caroline Ross

Purpose: To bring attention to nanoscale wall structures.

Suggestion: Delete unnecessary tick marks to unclutter both axes, and delete unnecessary parentheses (3.41a, 3.41b).

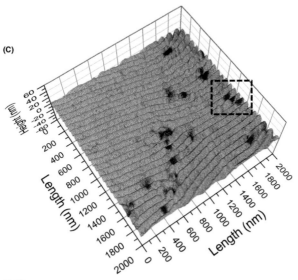

3.41a

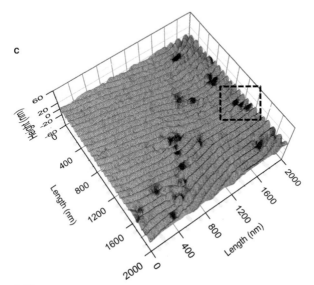

3.41b

Researcher: Connor Coley

Purpose: To show various components of an apparatus.

Suggestion: Delete unnecessary white boxes and simply label with text (3.42a, 3.42b).

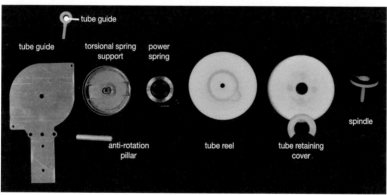

3.42a

3.42b

Researcher: Grant Knappe

Purpose: For a table of contents page submission (TOC)

Suggestion: Make labels readable and consider using all lower case (3.43a, 3.43b).

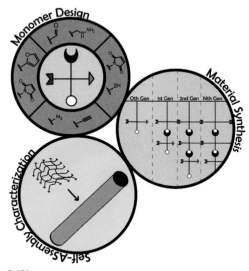

3.43a

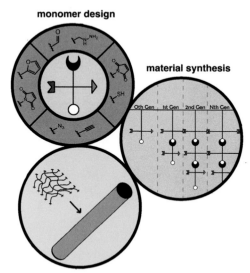

monomer design

material synthesis

self-assembly characterization

3.43b

Researcher: Kevin Lyons

Purpose: To compare NFL football players' performances using automated force-velocity profiling.

Suggestion: Kevin changed the position of each player's name, placing the labels close to the respective curve and in a matching color to improve visual clarity and make the legend less cluttered (3.44a, 3.44b).

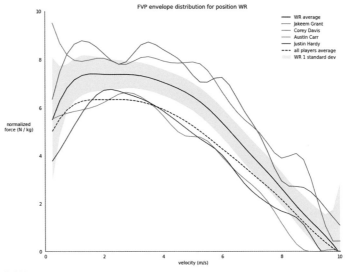

3.44a

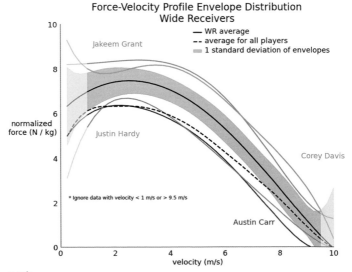

3.44b

Researcher: Weishan Liao

Purpose: To show parts of a Milwaukee Brewers pitching machine.

Suggestion: Straighten arrows and arrange labels in groups (3.45a, 3.45b).

Brewers Pitching Machine Visualization

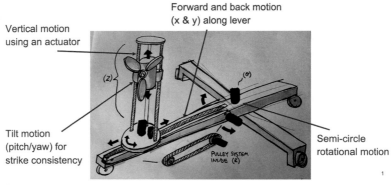

3.45a

Brewers Pitching Machine Visualization

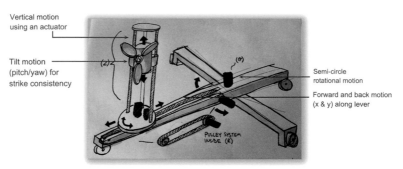

3.45b

Refining

The final strategy, refinement, is more about making small changes that include more than one of the previous strategies or fine-tuning your figure at the end of the process. It is here when you decide you have succeeded in making a clear and communicative figure and that you need to get some input from your colleagues, or even better, from someone outside your lab. Running a collaborative workshop in graphics should be part of your group's activities. I encourage your advisers to consider it. Participating in discussions specifically addressing visual communication helps you to think through your science.

BEFORE-AND-AFTER EXAMPLES

I am going to start with one of my own first redos. Years ago, I made two photographs of both sides of a long device in chemical engineer Klavs Jensen's lab. To make the first figure, the lab connected the two photographs and inserted fuchsia lines to indicate the complete actual device. The second figure is my revision.

Researcher: Klavs Jensen

Purpose: To show the structure of a microanalytic device.

Suggestions: Create more space around the whole image, delete arrows, use all lowercase letters, replace fuchsia with a subtler blue, and SEM inserts should match the orientation of the corresponding parts in the larger image (3.46a, 3.46b). Questions: Do we really need the inserts? Do they provide more information than the photographs alone?

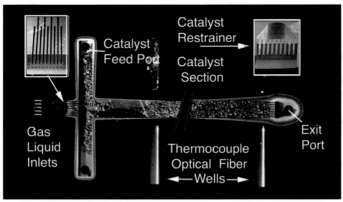

3.46a

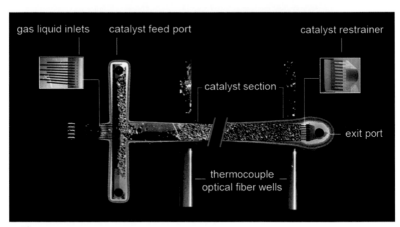

3.46b

Researcher: Ivan Susin Pires

Purpose: To illustrate how certain cytokines interact with key immune cells and to show how complex the interactions are.

Suggestions: Reorganize, delete unnecessary cartoons of cells, add drop shadows to bring attention to the cytokines convert all titles to black (3.47a, 3.47b).

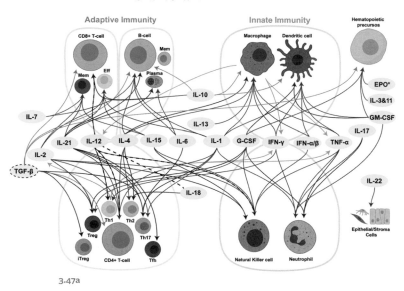

3.47a

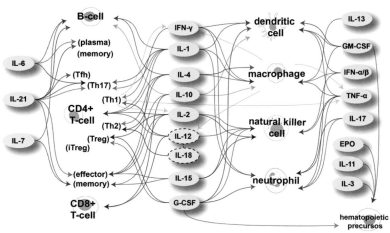

3.47b

Researcher: Joseph Maalouf

Purpose: To show the process of electrochemical lactonization.

Suggestions: Delete unnecessary blue background, and re-organize labels to improve overall composition (3.48a, 3.48b).

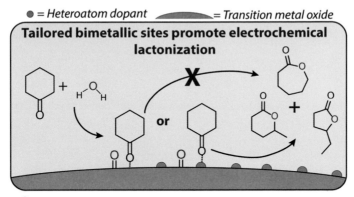

3.48a

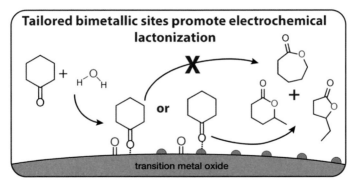

3.48b

Researcher: Abigail Frey

Purpose: To show the process of nanoparticle functionalization.

Suggestions: Abigail refined the first image by removing the circles representing phospholipid heads because they are not included in this system. She also lightened the gray centers, which seems to de-emphasize them relative to the other components (3.49a, 3.49b).

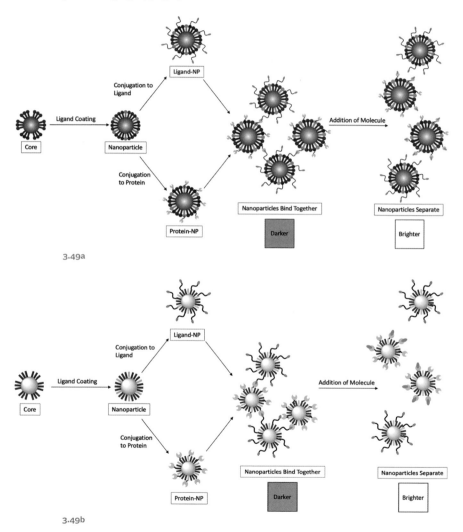

3.49a

3.49b

Researcher: Will Sawyer

Purpose: To show a process of nanoparticle formation to meet a journal's guidelines for a potential cover.

Suggestions: Delete the photograph and replace with an illustration, reorganize components to meet the requirements, and use drop shadows if they do help improve clarity (3.50a, 3.50b).

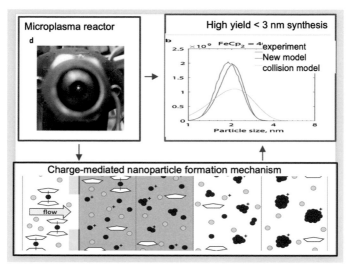

3.50a

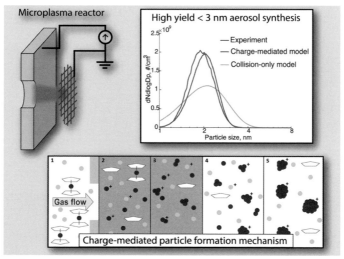

3.50b

Researchers: Ivan Jutamulia and Peko Hosoi

Purpose: To show where notable NBA teams fall on a two-dimensional plane of high-quality scoring opportunities and missed opportunity rate.

Suggestions: Ivan turned the bubbles into points, which removes the variance information but makes it easier to distinguish the teams. He also labeled the four quadrants and color-coded each team by quadrant (3.51a, 3.51b).

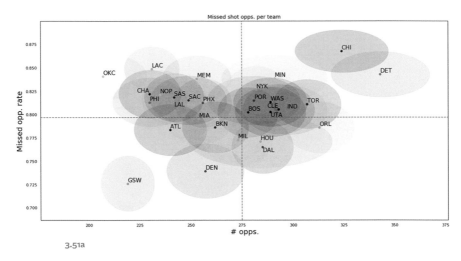

3.51a

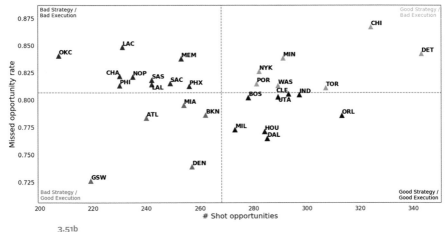

3.51b

Researcher: Katherine Mizrahi

Purpose: To show the structure of an important cell in a chemical engineering process.

Suggestions: Color the cell and the exploding diagram of it, delete color from other areas, and delete unnecessary frames (3.52a, 3.52b).

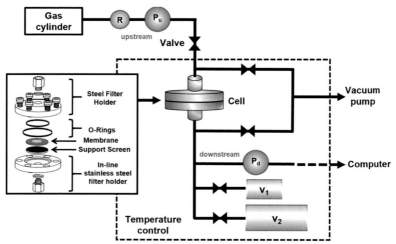

3.52a

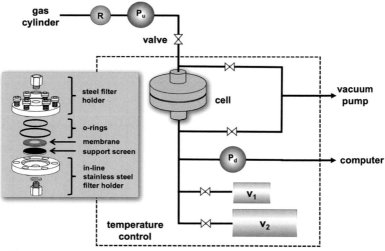

3.52b

Researchers: Krishna Shrinivas and B. Sabari

Purpose: To show the process of forming transcription-associated proteins in a graphical abstract.

Suggestions: For a graphical abstract, add a title at the top so that the viewer understands what the graphic is showing. Edit out redundant graphic pieces, rearrange pieces, and add a gray background so that the order of the story creates clarity (3.53a, 3.53b).

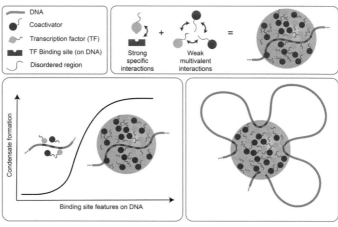

3.53a

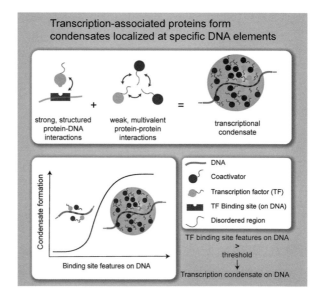

3.53b

Researcher: Miriam Gad

Purpose: To show the process of a graphene oxide sheet crumpling around silicon nanoparticles.

Suggestions: Miriam decided to enlarge the labels, label important arrows, and show more detail (3.54a, 3.54b).

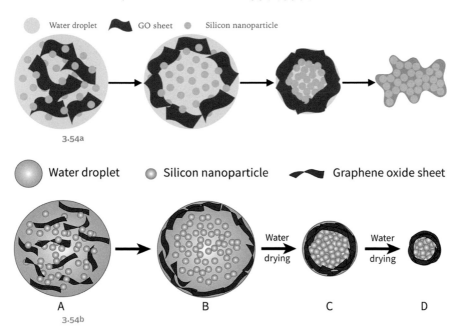

3.54a

3.54b

Researcher: Sahaofei Wang

Purpose: To show the process of creating vertically aligned channels from carbon nanotubes.

Suggestions: Add the graphene structure of the nanotubes (3.55a, 3.55b).

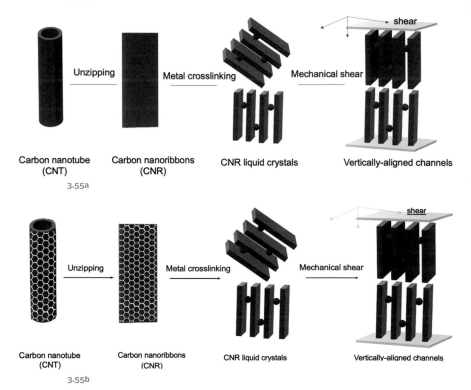

Carbon nanotube (CNT) — Unzipping — Carbon nanoribbons (CNR) — Metal crosslinking — CNR liquid crystals — Mechanical shear — Vertically-aligned channels — shear

3.55a

Carbon nanotube (CNT) — Unzipping — Carbon nanoribbons (CNR) — Metal crosslinking — CNR liquid crystals — Mechanical shear — Vertically-aligned channels — shear

3.55b

Generic system

Suggestion: Use thinner-lined frames and differentiate the inputs and outputs (ellipses) from the system (rectangles) (3.56a, 3.56b).

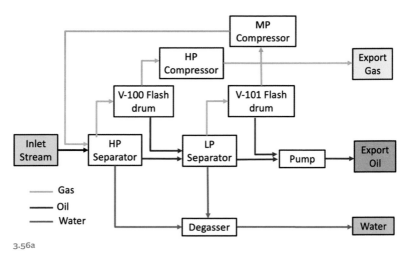

3.56a

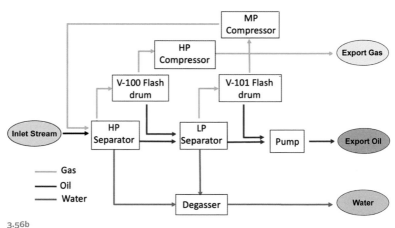

3.56b

Researcher: Wanny Woo

Purpose: To show the process of diffusion in the presence of a membrane.

Suggestions: Change the color of "chemical potential" so that it doesn't match the red particles. Add more particles on both sides of the membrane so the difference is easier to see. Connect all the colored lines to emphasize the change (3.57a, 3.57b).

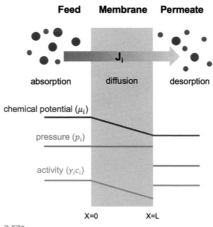

3.57a

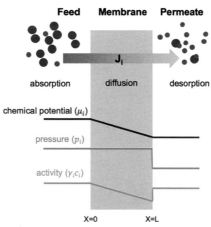

3.57b

Researcher: Hyunhee Lee

Purpose: To compare a proposed chemical mechanism to the "typical" mechanism.

Suggestion: Delete unnecessary elements, such as the callout frames and light blue backgrounds behind some labels (3.58a, 3.58b).

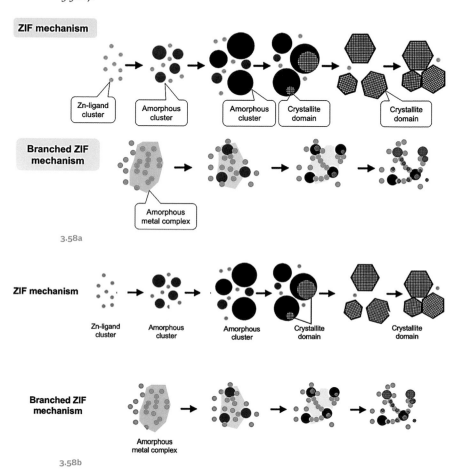

3.58a

3.58b

Note: The following is a before-and-after by Carin Cain, the illustration editor at Annual Reviews.

Purpose: To depict a tree, showing the basic principles underlying several designs for electrochemical nucleic acid sensors.

Carin's suggestions: (1) group more cohesively to establish a visual hierarchy, (2) use lowercase letters for the headings and roman

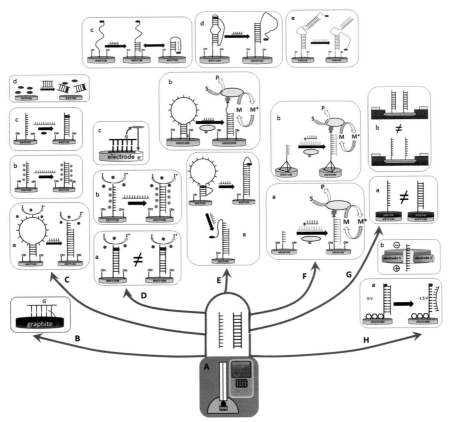

3.59a

numerals for the subheadings, (3) make each box a different color and color code the arrows to match the color of the box they point to, (4) enlarge the figure to be able to use a font no smaller than seven points while still adhering to the page-size limitations, and (5) change black arrows to AR's "house-style" arrows (3.59a, 3.59b).

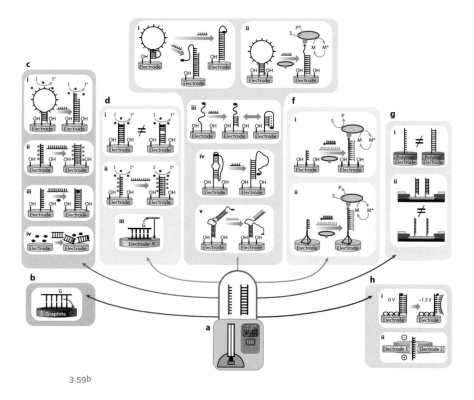

3.59b

We end with an exercise by graphic designer Alessandro Martemucci that brings us back to the idea that good design can aid our understanding not only of our science but also things we encounter in our everyday lives. The next time you buy a ticket—whether for a train, plane, or the theater—think about whether its design helps you understand. Alessandro is chief executive officer of the Officinae Lean Marketing Agency. He came up with his own redo of this Italian train ticket (3.60a, 3.60b). What do you think? This time, without my help, identify the changes he made and decide whether they work.

3.60a

3.60b

4

Posters and Slide Presentations

Posters

I once thought presenting a poster was a means of showing off your work to all participants at a conference, like in this photo by John Freidah of the MIT mechanical engineering grad student Cecile Chazot, who is explaining her research project (4.1). It turns out, at least in the minds of many MIT participants at my workshops, that poster sessions are intended to introduce your work to others who are interested in similar concepts, but not necessarily to all attending the conference. With that in mind, here are a number of quick ideas to think about designing your poster to help participants find you amid the sea of posters in a large hall.

4.1

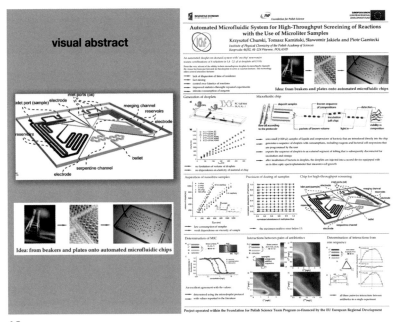

Automated Microfluidic System for High-Throughput Screening of Reactions with the Use of Microliter Samples

Krzysztof Churski, Tomasz Kamiński, Sławomir Jakiela and Piotr Garstecki

Institute of Physical Chemistry of the Polish Academy of Sciences

Kasprzaka 44/52, 01-224 Warsaw, POLAND

4.2

Edit! First and foremost, your poster is not a journal article. No one wants to read a full-length paper at a conference. Figure out what really needs to be shown and do it concisely.

Enlarge one or two images from your poster and display them at left with a colored background. If someone recognizes a device or a process from far away, they will be drawn to your poster to learn more (4.2).

Unfortunately, most conferences have strict formats that may not allow you to do these things. The Society of Photo-Optical Instrumentation Engineers, for example, insists that poster participants download a template that they are required to follow (4.3).

I'd say the society's template is well designed because of the simplicity and clarity of its sections, but I would be pretty disappointed walking into that hall of posters where all the posters would look the same. Wouldn't you?

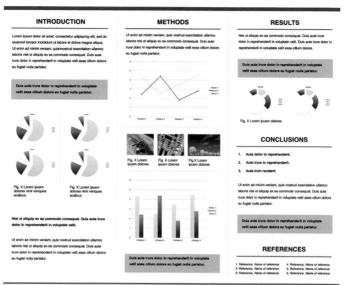

Poster Title Goes Here
Researchers'/Presenters' Names
Institution/Organization/Company

SPIE.

2ⁿᵈ LOGO

INTRODUCTION

Lorem ipsum dolor sit amet, consectetur adipiscing elit, sed do eiusmod tempor incididunt ut labore et dolore magna aliqua. Ut enim ad minim veniam, quisvnostrud exercitation ullamco laboris nisi ut aliquip ex ea commodo consequat. Duis aute irure dolor in reprehenderit in voluptate velit esse cillum dolore eu fugiat nulla pariatur.

Duis aute irure dolor in reprehenderit in voluptate velit esse cillum dolore eu fugiat nulla pariatur.

Fig. X Lorem ipsum dolores nimi venques ecellucs

Fig. X Lorem ipsum dolores nimi venques ecellucs

Nisl ut aliquip ex ea commodo consequat. Duis aute irure dolor in reprehenderit in voluptate velit.

Ut enim ad minim veniam, quis nostrud exercitation ullamco laboris nisi ut aliquip ex ea commodo consequat. Duis aute irure dolor in reprehenderit in voluptate velit esse cillum dolore eu fugiat nulla pariatur.

METHODS

Ut enim ad minim veniam, quis nostrud exercitation ullamco laboris nisi ut aliquip ex ea commodo consequat. Duis aute irure dolor in reprehenderit in voluptate velit esse cillum dolore eu fugiat nulla pariatur.

Fig. X Lorem ipsum dolores

Fig. X Lorem ipsum dolores

Fig.X Lorem ipsum dolores

Duis aute irure dolor in reprehenderit in voluptate velit esse cillum dolore eu fugiat nulla pariatur.

RESULTS

Nisi ut aliquip ex ea commodo consequat. Duis aute irure dolor in reprehenderit in voluptate velit. Duis aute irure dolor in reprehenderit in voluptate velit esse cillum dolore.

Duis aute irure dolor in reprehenderit in voluptate velit esse cillum dolore eu fugiat nulla pariatur.

Fig. X Lorem ipusm dolores

CONCLUSIONS

1. Aute dolor in reprehenderit.
2. Aute irure in reprehenderit.
3. Aute iruin renderit.

Ut enim ad minim veniam, quis nostrud exercitation ullamco laboris nisi ut aliquip ex ea commodo consequat. Duis aute irure dolor in reprehenderit in voluptate velit esse cillum dolore eu fugiat nulla pariatur.

Duis aute irure dolor in reprehenderit in voluptate velit esse cillum dolore eu fugiat nulla pariatur.

REFERENCES

1. Reference, Name of reference   4. Reference, Name of reference
2. Reference, Name of reference   5. Reference, Name of reference
3. Reference, Name of reference   6. Reference, Name of reference

Do not use a PowerPoint template. Unless you are required to use a template, why let someone else design your poster, especially when so many others will be using the same template? Your poster will become part of a uniform sea, making it harder to attract viewers.

Stick with one font, preferably sans serif. Keep in mind that using a sans serif is only an opinion. Studies are mixed about the ease of reading between serif versus sans serif. I prefer it because it's more of my aesthetic. You can, however, make some of the text bold, or use caps, or vary the size of the text as a means of creating a hierarchy of information. But adding more fonts only makes the poster too complicated.

Use italics minimally, except for the axes in a graph. Mixing text with unnecessary italics can be confusing.

Compartmentalize the sections with color. This directs viewers' eyes to what you want them to see and helps guide them through the poster, as the mechanical engineer Adam Paxson did (4.4). Notice the difference with and without color.

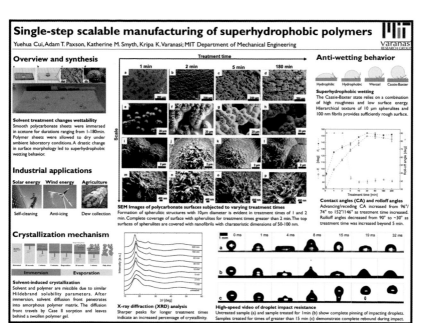

Single-step scalable manufacturing of superhydrophobic polymers

Yuehua Cui, Adam T. Paxson, Katherine M. Smyth, Kripa K. Varanasi; MIT Department of Mechanical Engineering

Overview and synthesis

Solvent treatment changes wettability
Smooth polycarbonate sheets were immersed in acetone for durations ranging from 1–180min. Polymer sheets were allowed to dry under ambient laboratory conditions. A drastic change in surface morphology led to superhydrophobic wetting behavior.

Industrial applications

Solar energy   Wind energy   Agriculture

Self-cleaning   Anti-icing   Dew collection

SEM Images of polycarbonate surfaces subjected to varying treatment times
Formation of spherulitic structures with 10µm diameter is evident in treatment times of 1 and 2 min. Complete coverage of surface with spherulites for treatment times greater than 2 min. The top surfaces of spherulites are covered with nanofibrils with characteristic dimensions of 50-100 nm.

Anti-wetting behavior

Hydrophilic   Hydrophobic   Wenzel   Cassie-Baxter

Superhydrophobic wetting
The Cassie-Baxter state relies on a combination of high roughness and low surface energy. Hierarchical texture of 10 µm spherulites and 100 nm fibrils provides sufficiently rough surface.

Contact angles (CA) and rolloff angles
Advancing/receding CA increased from 96°/74° to 152°/146° as treatment time increased. Rolloff angles decreased from 90° to ~30° as treatment time was increased beyond 5 min.

Crystallization mechanism

Immersion   Evaporation

Solvent-induced crystallization
Solvent and polymer are miscible due to similar Hildebrand solubility parameters. After immersion, solvent diffusion front penetrates into amorphous polymer matrix. The diffusion front travels by Case II sorption and leaves behind a swollen polymer gel.

X-ray diffraction (XRD) analysis
Sharper peaks for longer treatment times indicate an increased percentage of crystallinity.

High-speed video of droplet impact resistance
Untreated sample (a) and sample treated for 1min (b) show complete pinning of impacting droplets. Samples treated for times of greater than 5 min (c) demonstrate complete rebound during impact.

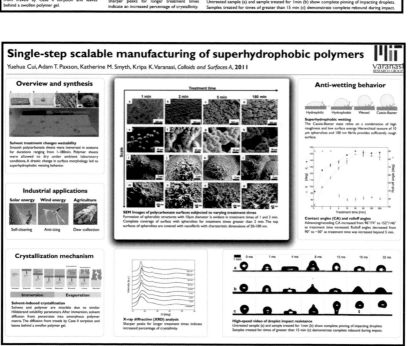

Single-step scalable manufacturing of superhydrophobic polymers

Yuehua Cui, Adam T. Paxson, Katherine M. Smyth, Kripa K. Varanasi, *Colloids and Surfaces A*, 2011

Overview and synthesis

Solvent treatment changes wettability
Smooth polycarbonate sheets were immersed in acetone for durations ranging from 1–180min. Polymer sheets were allowed to dry under ambient laboratory conditions. A drastic change in surface morphology led to superhydrophobic wetting behavior.

Industrial applications

Solar energy   Wind energy   Agriculture

Self-cleaning   Anti-icing   Dew collection

SEM Images of polycarbonate surfaces subjected to varying treatment times
Formation of spherulitic structures with 10µm diameter is evident in treatment times of 1 and 2 min. Complete coverage of surface with spherulites for treatment times greater than 2 min. The top surfaces of spherulites are covered with nanofibrils with characteristic dimensions of 50-100 nm.

Anti-wetting behavior

Hydrophilic   Hydrophobic   Wenzel   Cassie-Baxter

Superhydrophobic wetting
The Cassie-Baxter state relies on a combination of high roughness and low surface energy. Hierarchical texture of 10 µm spherulites and 100 nm fibrils provides sufficiently rough surface.

Contact angles (CA) and rolloff angles
Advancing/receding CA increased from 96°/74° to 152°/146° as treatment time increased. Rolloff angles decreased from 90° to ~30° as treatment time was increased beyond 5 min.

Crystallization mechanism

Immersion   Evaporation

Solvent-induced crystallization
Solvent and polymer are miscible due to similar Hildebrand solubility parameters. After immersion, solvent diffusion front penetrates into amorphous polymer matrix. The diffusion front travels by Case II sorption and leaves behind a swollen polymer gel.

X-ray diffraction (XRD) analysis
Sharper peaks for longer treatment times indicate increased percentage of crystallinity.

High-speed video of droplet impact resistance
Untreated sample (a) and sample treated for 1min (b) show complete pinning of impacting droplets. Samples treated for times of greater than 15 min (c) demonstrate complete rebound during impact.

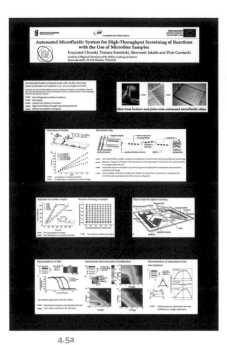

4.5a

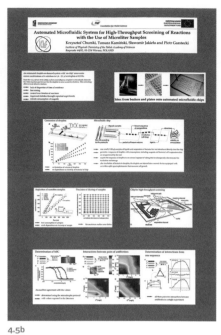

4.5b

Do not use strong colors for backgrounds. Stick with gray for posters
with this design. That way, no color in the poster has to fight with
another color (4.5a, 4.5b).

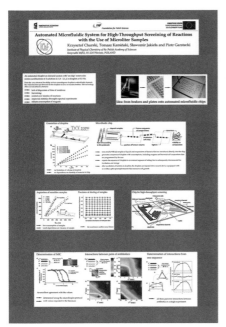

4.6a

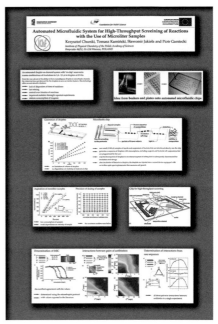

4.6b

Try using drop shadows to make each section pop out (4.6a, 4.6b).

Avoid using colored titles or underlines for subsections. These unnecessarily add graphical distractions.

Use bullets, not arrows. Stylized arrows add graphical distractions.

Create more space around the perimeter of each segment. White space improves readability.

Why Nonlinear Imaging?

- Our method encodes information about what happens in the wavelength of the light - even scattered light can give useful information.
- Two Photon Absorption (TPA), and Self Phase Modulation (SPM) both give rise to reshaping a spectral pulse (see below).
- We use a lock-in amplifier to selectively look at one section of the frequency spectrum, giving a <u>virtually background free measurement</u>.

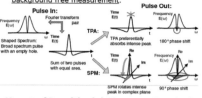

How to (Sensitively) Measure Spectral Changes:
Homodyne Measurement

- Use a 4-f pulse shaper to create a spectral hole.
  - In AOM plane, position is mapped to frequency.
  - Intensity & phase of RF signal modulates intensity & phase of diffracted light.

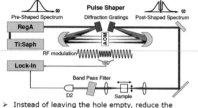

- Instead of leaving the hole empty, reduce the amplitude, and rotate the phase.
- Lock-in to this phase rotation:
  - In phase refilling: TPA
  - Out of phase refilling: SPM

4.7a

Why non-linear imaging?

- Our method encodes information about what happens in the wavelength of the light - even scattered light can give useful information.
- Two Photon Absorption (TPA), and Self Phase Modulation (SPM) both give rise to reshaping a spectral pulse (see below).
- We use a lock-in amplifier to selectively look at one section of the frequency spectrum, giving a <u>virtually background free measurement</u>.

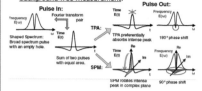

Homodyne measurement and spectral changes

- Use a 4-f pulse shaper to create a spectral hole.
  - In AOM plane, position is mapped to frequency.
  - Intensity & phase of RF signal modulates intensity & phase of diffracted light.

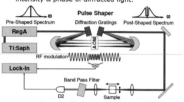

- Instead of leaving the hole empty, reduce the amplitude, and rotate the phase.
- Lock-in to this phase rotation:
  - In phase refilling: TPA
  - Out of phase refilling: SPM

4.7b

The last three suggestions can all be seen in my revision of a segment of another poster. Note that I shortened the lower title text in the "after" version (4.7a, 4.7b).

There are many resources for poster presentation today, and a few good starting points are included here. Some of these go into more than just design, including subjects such as how you should dress, about which I am the last person to make suggestions.

As a few examples, I suggest looking online for tips from photographer and biologist Colin Purrington (https://colinpurrington.com/tips/poster-design/), at a *Nature* article on conference presentations (https://www.nature.com/articles/nj7614-115a), and at a *Science Careers* article with advice on poster presentations (https://www.science.org/content/article/how-prepare-scientific-poster).

I'll end this poster discussion with something to think about as technology continues to develop, and that is the prospect of interactivity. I am convinced the whole enterprise of poster presentation and design will change drastically in the near future, at least assuming that those who make conference decisions keep an open mind.

Mike Morrison, for example, has an entertaining and popular YouTube video in which he encourages scientists to make posters that prominently feature a QR code (4.8). He proposes restricting

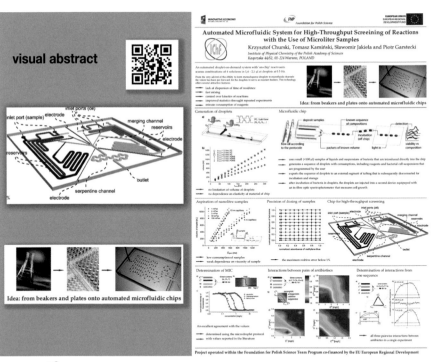

4.8

the usual sections (Intro, Methods, Results) to the sides of the poster and reserving the center for a lay-language title and a large QR code that directs viewers to a complete description of the research (you can watch his video "How to Create a Better Research Poster in Less Time" at his YouTube page).

Mike is wonderfully communicative and has put together some thoughtful ideas, but imagine coming to the entrance of the poster hall and seeing a sea of these poster designs. Would you be drawn to one particular QR code, or would the attraction work better with a stunning and informative image of the presenter's research or equipment? His second iteration does feature images from the research and makes the QR codes much smaller (https://youtu.be/SYk29tnxASs?t=1180).

To my mind, the interactive piece is what's most important here—and so much more will be developed. Think about how a viewer could see an animation or a relevant video right on your poster or, for that matter, how the entire poster could be interactive. I find it all very exciting.

Slide Presentations

When was the last time you viewed your presentation from the audience's point of view? Imagine walking into an auditorium or conference room to see what a viewer sees while you remotely advance your slides.

You might be surprised to realize that your text is too small, your slides are too busy, you cannot figure out where to look first, and you feel bombarded with colors. That's just for starters.

I also suggest looking at the slide presentations when presented by others and considering what annoys you. What's frustrating? It's a means of learning from others' mistakes.

More important, what design decisions have the presenter made that help to communicate? Think about using them.

Here is a list of suggestions to consider when you put together your talk, some of which echo—and I hope reinforce—suggestions I've made elsewhere in this volume:

- Try not to be seduced into using a PowerPoint template, even if you think it will save time. Your presentation is an opportunity to tell your story in your own voice and stand out from all the others. That means your slides should be designed by you.
- Reduce clutter. Think about editing the text down to the bare minimum for each slide.
- Make sure all slides have a consistent look.
- If you must show your school logo, place it in the lower right corner and make it grayscale. That too will help reduce visual clutter.
- Animate your slides: start with a simple canvas with one idea and add more information bit by bit as you advance.
- Use only one font to simplify reading the text (4.9, 4.10). I prefer Helvetica Neue, but that is only my opinion. Stay with one font of your choice. You might first like the first one on the next page because it looks fun, but fonts that call attention to themselves only distract the audience from your research. The simpler, the better.

USING **TEXT** IN A *SLIDE* <u>PRESENTATION</u>	Using Text in a Slide Presentation

4.9   4.10

- Consider dark backgrounds to make text and images pop out, and use color sparingly for text. Keep in mind that dark backgrounds work well on a screen but not on the printed page (4.11).

4.11

- Try using drop shadows for texts if it helps the viewer to read the text (4.12a, 4.12b). Test it out with your colleagues. Look carefully at the two images and note the difference drop shadows make in the image on the right. In my opinion, the text is easier to read with shadows, but make sure you don't overdo their use.

4.12a   4.12b

- The text of the labels and headers on your slide should be identical to the words you are speaking in your presentation. This is something that has always driven me crazy. If you are saying one thing, and the viewer is trying to read different words on your slide, it's much harder for her to follow.
- Don't underline your text. It is not necessary. Use various sizes of the same font or bold font or capital letters to create a hierarchy of information for headings and subheadings.
- Use a simple sentence or even ask a question for your opening slide. Computer scientist Daniel Jackson simplifies the text, which makes a powerful impression (4.13).

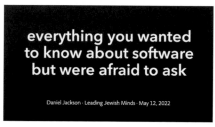

4.13

Finding Your Own Voice

This is a good place to discuss your visual "voice" because the discussion is relevant to the previous image. Can you find an aesthetic in designing your presentations that is *yours*? Daniel Jackson's visual "voice" is one of elegance and simplicity. In a recent Zoom talk, he started with the intro slide and maintained a consistent visual style throughout his presentation (4.14–4.19). Here are some of the things he did to accomplish this (see the next series of slides):

the herman illusion: perception & brain function

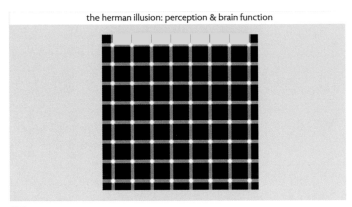

4.14

windows vs. mac: which is better?

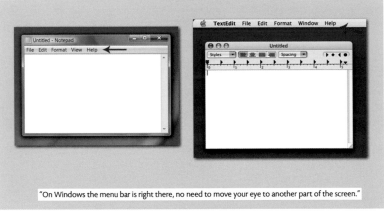

"On Windows the menu bar is right there, no need to move your eye to another part of the screen."

4.15

five dark concepts of surveillance capitalism

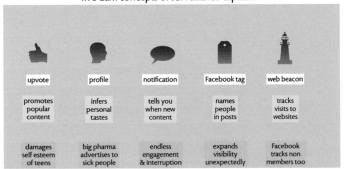

upvote	profile	notification	Facebook tag	web beacon
promotes popular content	infers personal tastes	tells you when new content	names people in posts	tracks visits to websites
damages self esteem of teens	big pharma advertises to sick people	endless engagement & interruption	expands visibility unexpectedly	Facebook tracks non members too

4.16

henry dreyfuss, designing for people (1955)

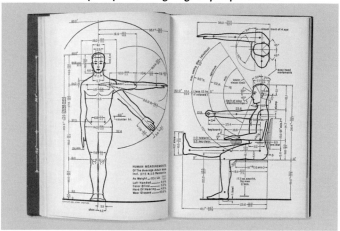

4.17

4.18

closing thoughts

4.19

- Used only a sans serif font without capital letters for slide titles
- Used a light-gray background to help the colored images pop
- Used a minimal amount of text, especially on slides marking a transition in his talk
- Started each slide speaking only the words on the screen to avoid confusing the audience

Consider another different example of a presentation design (4.20a, 4.20b). Elba Alonso-Monsalve is a graduate student in physics. Look at a few slides from her talk for the general public. Note how she "animates" the slides by adding information on each slide and moves along the talk (4.21a, 4.21b).

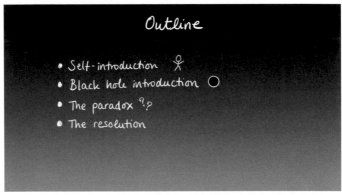

4.20a

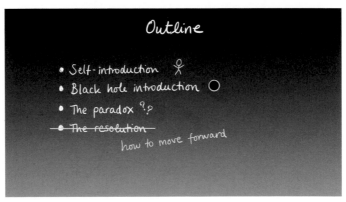

4.20b

4.21a

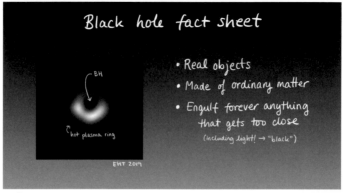

4.21b

Notice her aesthetic. It might not be the right choice for a talk to her scientific colleagues, but I would guess the audience was keen to pay attention. Elba used Notability on her iPad to hand-draw these slides. It's an app often used to annotate papers. She started by making one slide on her computer using Google Slides, with just the background color gradient and the desired length-width ratio. Then she saved the slide as a PDF and imported it into Notability. She didn't use any fonts; that's just her handwriting.

When I first saw Juliann Tefft's illustrations in a keynote talk given by Chris Chen for the Biomedical Engineering Society at Boston University, I just knew I had to contact her. The work was so different from other illustrations I had seen in talks. It was clearly done with a human hand, with a refined aesthetic, and most of all, was wonderfully communicative. She first sketched out a vision of the figure and discussed it with Chris (4.22).

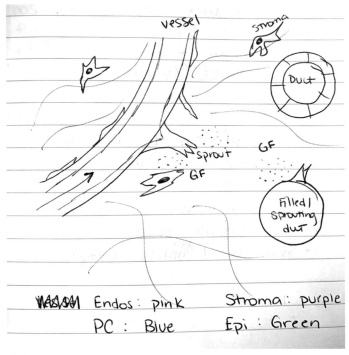

4.22

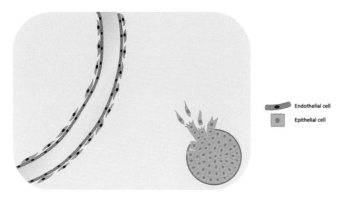

Blood vessels interact with surrounding tissues through chemical and mechanical signaling.

4.23a

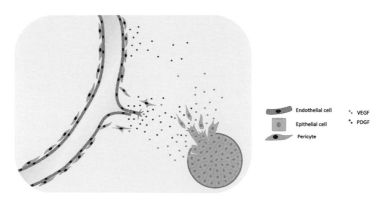

This stimulates the blood vessel nearby to extend a new vessel sprout (red cells) towards the tumor. The endothelial cells lining this sprouting blood vessel secrete the factor PDGF (red dots). During this process, pericytes (blue cells) dissociate from the vessel but are attracted back towards the developing sprout by PDGF.

4.23b

Her edited set of slides shows the process of a blood vessel interacting with cancer cells as an example of how cell behavior—such as the growth of a blood vessel toward a tumor, which depends on both chemical and mechanical signals (4.23a–4.23d).

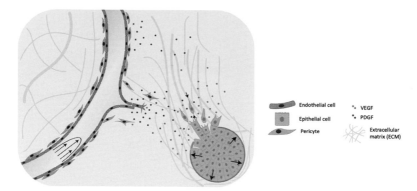

Within the cancerous epithelial duct (green), dysregulated cell proliferation causes the cells to exert pressure on the tissue surrounding the tumor and alters the matrix alignment (tan fibers) in the surrounding tissue.

4.23c

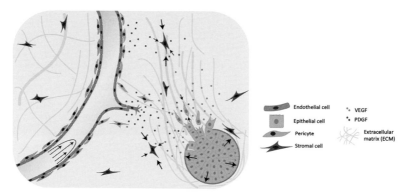

Aligned matrix stimulates cells to travel along these fibers, such as the invading epithelial cells (green) and surrounding stromal cells (purple). These forces are felt by cells nearby, again stimulating dynamic changes in their neighbor's behavior.

4.23d

Because the presentation slides were to move quite quickly, she refined the schematic over time to help the audience grasp the concepts as intuitively as possible. From the start, she wanted each cell type to be a unique color to intuitively signal that they are different. The precise colors were edited over time. The blood vessel, for instance, was initially a lighter pink, but through feedback from Chris, she made this a brighter red to improve contrast with the background.

She wanted the endothelial cells lining the blood vessel to be a reddish color since blood is red. For the other colors, she just wanted to maximize contrast! The growth factors did change color through the process. Initially, she showed one set as red and one set as black, but through the refining process, she realized that the connection of the growth factors to the different cell types could be better captured if the colors were consistent with the cell of origin.

Juliann and I discussed how powerful this storyboard would be in animation. Remember, an animation is a series of still images. In presenting these, clicking through the slides quickly could have worked in the same way.

Mohammed AlQuraishi studies machine-learning models for predicting protein structure and function and protein-ligand interactions. The material in his talk was very complicated, but Mohammed developed a remarkable visual vocabulary and color palette to clearly communicate the work. He started with a simple introductory slide to explain his conceptual system for the talk (4.24a).

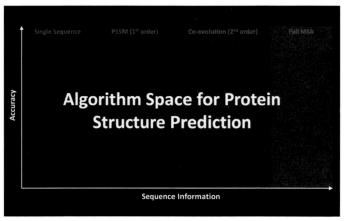

4.24a

He then added information to that template in the slides that followed (4.24b, 4.24c). Getting from the first to the last took about eight slides.

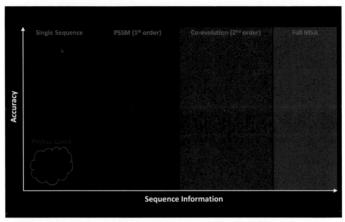

4.24b

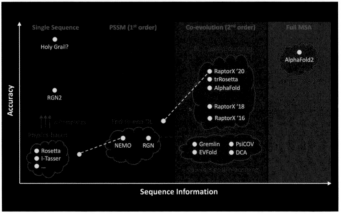

4.24c

Because of the complexity of the subject, his summary slide was included in the upper right corner in the rest of his slides to give context to the process (4.24d, 4.24e). I wish more people thought about doing that. Sometimes the talks are so dense that it would be nice to have a signal indicating "you are here."

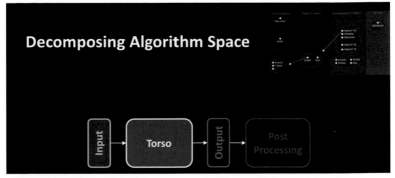

4.24d

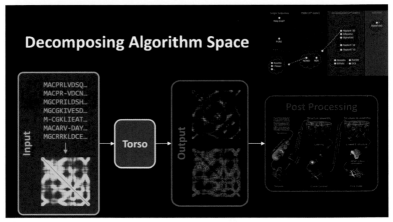

4.24e

Mohammed's use of color is special, and it works. Notice how he is not giving us solid blocks of color. That would be too much to look at. Instead, he suggests color with translucent blocks, using darker frames in the same color to create sections. Also note that the titles are white, which makes them stand out. All work beautifully together to make a complicated yet cohesive visual story with a highly successful personal voice.

A Last Word or Two

Do you feel bombarded with new ideas? Good! If you skipped examples that weren't related to your research subject, please go back and review them. No matter your discipline, I am convinced that each of these examples can help you represent your work. Let yourself think deeply about how you design your visuals. Remember to simplify and keep in mind who you are writing and speaking to. Experiment to see what works and ask around to hear if others agree. Good design decisions will improve your communication and help you and others understand the critical and fascinating research world you made a conscious decision to join.

When I first started at MIT, my work was mostly about photographing the institute's research. It is still a major part of what I do, but I later realized there was another need in young researchers' education. With our workshops and one-on-one conversations, students, postdocs, and even established principal investigators are seeing the value of design when creating their informational graphics for publication. Today, when I ask the various groups what they look at when reading an article, they invariably say, in the following order, title, abstract, and then right to the graphics. If graphics play such a critical role in publishing, why is their design left for the last minute, and why are they not taken more seriously? Unfortunately, education in that area is almost nonexistent on most university campuses. These handbooks, at the very least, I hope will start the needed conversations.

But none of these handbooks would have been possible without the hundreds of remarkable students and faculty, mostly at MIT, with whom I have had the privilege of working. As I previously noted, the best part of my professional life is that every conversation brings yet another piece of fascinating science into my life. I am grateful to them all.

As for the book, my gratitude goes to Joe Calamia, my editor at the University of Chicago Press, who continues to keep me laughing, calm, and on time. Greg Miller did an extraordinary job editing my first draft. And my deepest thanks go to Isaac Tobin, whose talent met the challenges of designing a small handbook with precision and intelligence.

And most of all, my love and gratitude go to my extraordinary family and to my dearest friends. All know who they are.

Credits

0.2.	Reproduced with permission from the Boston Athletics Association.
1.1.	Casa Buonarroti
1.4.	Royal Collection Trust / © His Majesty King Charles III 2023.
1.6a–b.	The National Portrait Gallery. *National Gallery Technical Bulletin. 'Coloured' Infrared Image Portrait of Giovanni(?) Arnolfini and his Wife. Jan van Eyck.*
1.13b.	Notebook B (DAR 121). Reproduced by kind permission of the Syndics of Cambridge University Library.
2.1.	Cahill-Keyes projection by Gene Keyes.
2.6.	Z. Jane Wang, Raymond Chang, and Leif Ristroph, *Dragonfly Righting Reflex*, APS-DFD Video, 2021. https://gfm.aps.org/meetings/dfd-2021/6137b171199e4c7029f44abe.
2.7b.	1972 NYC Subway Map © Metropolitan Transportation Authority. Used with permission.
2.8.	Reproduced with permission from *Annual Reviews*. K. L. Law, *Annual Review of Marine Science* 9 (2017): 205–29, https://doi.org/10.1146/annurev-marine-010816-060409.
2.10.	Infographic from the article "The Race for Coronavirus Vaccines: A Graphical Guide," by Ewen Callaway, *Nature*, April 28, 2020, https://www.nature.com/articles/d41586-020-01221-y.
3.1.	Credit: Trent Schindler/NASA.
3.2.	Credit: Matthew Glasser and David Van Essen.
3.20a–b.	Reproduced with permission from *Annual Review of Fluid Mechanics*, Volume 36, 2004 © by Annual Reviews, http://www.annualreviews.org.
3.35.	Credit: Jay M. Pasachoff and Alex Filippenko, *The Cosmos: Astronomy for the New Millennium*, 5th ed. (New York: Cambridge University Press).
3.59a–b.	Used by permission. Created by Carin Cain, from Elena E. Ferapontova, "DNA Electrochemistry and Electrochemical Sensors for Nucleic Acids," *Annual Review of Analytical Chemistry* 11 (2018): 197–218, https://doi.org/10.1146/annurev-anchem-061417-125811.